天津大学出版社
TIANJIN UNIVERSITY PRESS

international landscape wind vane

国际景观风向标

李壮 编

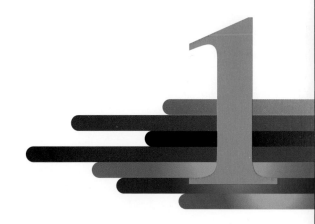

天津大学出版社
TIANJIN UNIVERSITY PRESS

图书在版编目 （CIP）数据

　　国际景观风向标．1／李壮编．—天津：天津大学
出版社，2013.3
　　ISBN 978-7-5618-4621-6

　　Ⅰ．①国… Ⅱ．①李… Ⅲ．①景观设计—作品集—世
界—现代Ⅳ．① TU968.2
　　中国版本图书馆 CIP 数据核字（2013）第 031378 号

主　编：李壮

责任编辑：朱玉红

执行主编：石晓艳

编　委：李秀　刘云　韦成刚　高松　赵睿

翻　译：国歌　樊喜强　王艳秋

编　辑：肖娟

设　计：⑦ 北京吉典博图文化传播有限公司

出版发行：天津大学出版社

出版人：杨欢

地　址：天津市卫津路 92 号天津大学内（邮编：300072）

电　话：发行部 022-27403647　邮购部 022-27402742

网　址：publish. tju. edu. .cn

印　刷：上海锦良印刷厂

经　销：全国各地新华书店

开　本：220mm×300mm

印　张：42

字　数：785 千

版　次：2013 年 4 月第 1 版

印　次：2013 年 4 月第 1 次

定　价：736.00 元（全两册）

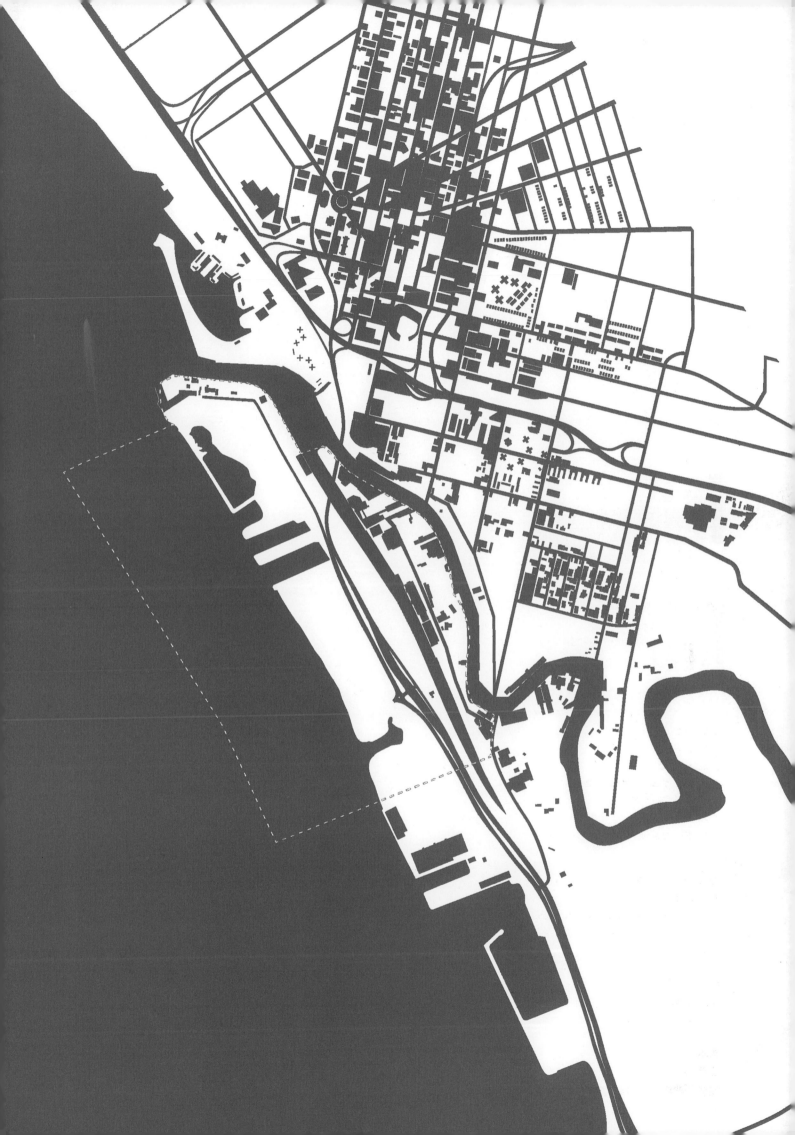

CONTENTS

Pok Kobkongsanti, MLAUD
Projects

Mandarin Oriental Resort and Spa, Hainan, China
Keyne by Sansiri, Bangkok, Thailand
PYNE by Sansiri, Bangkok, Thailand
Marriott Hotel , Xiangshui Bay, Sanya, China
M Phayathai, Bangkok, Thailand
HSBC Green Library, Bangkok, Thailand
J Residence, Bangkok, Thailand
2009
M Silom, Bangkok, Thailand
Central Plaza Ladphrao Renovation, Bangkok, Thailand
Central Plaza Rama 9, Bangkok, Thailand
Suzhou Residences by KWG,Phase 1-4, Suzhou, China
Marriott Boutique Hotel, Sanya, China
Hideaway Hotel Chibi, China
2008 Hilton Hotel Pattaya, Chonburi, Thailand
Baan Sansuk by Sansiri, Hua Hin, Thailand
Quattro by Sansiri, Bangkok, Thailand
Zense Gourmet Deck and Lounge Panorama, Bangkok, Thailand
Pinnacle Collection, High-end Residential Project , Sentosa, Singapore
Noble Remix, Bangkok, Thailand
Radisson Maldives, Maldives
Jodhpur Hotel, Jodhur, India
Sense Sukhumvit, Bangkok, Thailand
2007
Central Plaza Chonburi, Chonburi, Thailand
Noble Nano, Bangkok, Thailand
Noble Cube, Bangkok, Thailand
Noble Residence, Bangkok, Thailand
Seascape, High-end Residence, Sentosa, Singapore
Hotel Lotus, Hailing, China
Prive' by Sansiri, High-end Residence, Bangkok, Thailand
Preen by Sansiri, High-end Residence, Bangkok, Thailand
Avora31, High-end Service Apartment, Sukhumvit 31, Bangkok, Thailand
2005—2006
W Hotel, Guangzhou, China
The Four Seasons Resort and Spa, West Lake, Hangzhou, China
Danang Sontra Resort and Spa, Danang, Vietnam
One and Only Hotel, Cape Town, South Africa
Park Hyatt Hotel, Ningbo, China
Coral Island, Luxury Seaside Villas , Sentosa, Singapore
Cosmos Luxury Highrise Residence, High-end Residential Project , Guangzhou, China
Phoenix City, High-end Residential Project , Beijing, China
2000—2004
World Trade Center Master Plan, New York, USA
Design Guideline for the Overall Appearance of the New World Trade Center
21st Century Waterfront Park, Chattanooga, Tennessee, USA
Waterfront Development Study Prepared for the City's 15 Years Plan
Louisville Waterfront Development, Louisville, USA
The Second Phase of the Award-winning Public Openspace
Clinton Presidential Center, Little Rock, Arkansas, Kentucky, USA
New Public Park and Library for the City of Little Rock, Arkansas, USA
Main Street, University of Cincinnati, Ohio, USA
The Main Openspace of the University

Pok Kobkongsanti

Pok Kobkongsanti was graduated from Graduate School of Design, Harvard University, in 2000. Since then, he had practiced with Mr. George Hargreaves of Hargreaves Associates and Mr. Bill Bensley of Bensley Design Studio.
In 2007, he found T.R.O.P : terrains + open space, with hope to develop a new ground in landscape architectural design.
Currently, Pok and T.R.O.P have been working on projects throughout Asia Region.

For us, each project deserves its own identity.

T.R.O.P

T.R.O.P

T.R.O.P is an initial of terrains and open space. We are a landscape architectural design studio, with a team of designers and construction supervisors.

Led by Pok Kobkongsanti, who brings in years of experience in USA and Asia, our philosophy is to create a unique design for each project. We work closely with our clients during the design process to make sure that each project gets its own definition of the landscape architecture. For us, our design process is as important as the design itself.

Together with our clients, we have successfully defined landscape architectures that can respond to requirements of their users. Since 2007, T.R.O.P has been working on various kinds of projects throughout Asia Region.

Currently our works include hospitality, high-end residential, commercial and installation designs. Each of our works has its own character, which is different from one another.

Baan Sansuk

Baan Sansuk 民居

LOCATION：Thailand
项目地点：泰国

AREA：11,613m²
面积：11 613 平方米

COMPLETION DATE：2010
完成时间：2010 年

PHOTOGRAPHER：Pattarapol Jormkhanngen
摄影师：Pattarapol Jormkhanngen

DESIGN DIRECTOR：Pok Kobkongsanti
设计总监：Pok Kobkongsanti

TEAM：Wasin Muneepeerakul，Pakawat Varaphakdi，Kampon Prakobsajakul
团队：Wasin Muneepeerakul，Pakawat Varaphakdi，Kampon Prakobsajakul

DESIGN COMPANY：T.R.O.P
设计公司：T.R.O.P

Baan Sansuk

Baan Sansuk is an exclusive residential project, located at Hua Hin, Thailand's most popular beach. The site is long, noodle—like with a small narrow side connected to the beach. There are two rows of buildings on both sides, leaving a long space in the middle of the site. Basically, most of the units, except the beach—front ones, do not have any ocean view. Instead, they are facing the opposite units.

Baan Sansuk 是一个独一无二的民居项目，坐落于泰国非常受欢迎的海滩——华欣。场地是长条形的，就像面条一样，小而狭窄的一边连接着海滩。在场地的两边分别排列着两排建筑，中间则留有长条形的空地。基本上，大多数的单元住宅，除了靠近海滩的那些之外，并不能看到海洋风光。相反，它们朝向对面的单元房间。

Our first move is to bring "the view" into the residence instead. Our inspiration of "the view" comes from the location of the project. Hua Hin, in Thai, means stone. The name comes from the natural stone boulders in its beach area. So we proposed a series of swimming pools from the lobby to the beach area, a total length of 230 m.

我们行动的第一步是将"风景"引入住宅区。我们关于"风景"设计的灵感来源于这个项目的所在地。华欣，在泰语里意为石头。这个名字来源于坐落在华欣海滩区域的自然巨石群。所以，我们提议建设一系列的游泳池，从场所的大厅一直延伸到海滩区域，总共长 230 米。

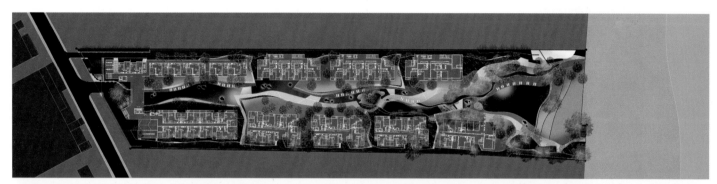

The pools are divided into several parts, with different functions like Reflecting Pool, Kids Pool, Transitional Pool, Jacuzzi Pool and Main Pool. At some certain area, we strategically place natural stone boulders to mimic the famous local beach. The result is a breath—taking water landscape, with different water characters from one end of the site to another. These pools are not just for eye—pleasure, but they also serve as the pools for everyone in the family.

游泳池按其不同的功能被分为几大部分，如将神殿倒映于水中的石砌水池、儿童游泳池、过渡游泳池、配有冲浪按摩浴缸的游泳池和主游泳池。在某些特定的区域，我们策略性地放置了自然石群去模仿当地著名的海滩景观。我们的设计带来了令人惊叹的水体景观效果，不同的水体特征从场地的一端变化到另一端。设置这些游泳池的目的不仅是为了带来视觉上的愉悦，同时，它们还可服务于家庭中的每位成员。

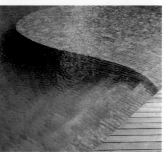

Casa De La Flora

卡萨德拉弗罗兰

LOCATION：Thailand
项目地点：泰国

AREA：8,750 m²
面积：8 750 平方米

COMPLETION DATE：2011
完成时间：2011 年

PHOTOGRAPHER：Pok Kobkongsanti
摄影师：Pok Kobkongsanti

DESIGN DIRECTOR：Pok Kobkongsanti
设计总监：Pok Kobkongsanti

TEAM：Kampon Prakobsajakul，Teerayut Pruekpanasan
团队：Kampon Prakobsajakul，Teerayut Pruekpanasan

DESIGN COMPANY：T.R.O.P
设计公司：T.R.O.P

Casa De La Flora

Hilton Central Pattaya Hotel

希尔顿中心芭堤雅酒店

LOCATION：Thailand
项目地点：泰国

AREA：1st floor ＝1,252 m² 16th floor ＝2,538 m² 16Mth ＝408.51 m²
面积：1 楼 1 252 平方米 16 楼 2 538 平方米 16 楼夹层 408.51 平方米

COMPLETION DATE：2009
完成时间：2009 年

PHOTOGRAPHER：Adam Brozzone，Charkhrit Chartarsa
摄影师：Adam Brozzone，Charkhrit Chartarsa

DESIGN DIRECTOR：Pok Kobkongsanti
设计总监：Pok Kobkongsanti

TEAM：Wasin Muneepeerakul，Pattarapol Jormkhanngen，Teerayut Pruekpanasan，Chatchawan Banjongsiri
团队：Wasin Muneepeerakul，Pattarapol Jormkhanngen，Teerayut Pruekpanasan，Chatchawan Banjongsiri

DESIGN COMPANY：T.R.O.P
设计公司：T.R.O.P

Hilton Central Pattaya Hotel

When we first visited the site, we noticed three important elements.
1.A gigantic skylight in the middle of the roof.
This skylight is to bring sunlight down to the mall.
It can't bear any load on it. So we can't use it as part of the garden.
To make it worse, the mall didn't have budget to decorate this skylight at all, so it is half glass and half concrete, and we ended up putting fake grass on it.

第一次到达场地时，我们注意到了三个重要的元素。
第一，屋顶中间的巨大天窗。
阳光可以透过天窗照进购物中心。
这个天窗不能承重，所以我们不能将其用做花园的一部分。
更糟糕的是，购物中心根本就没有可用于天窗装饰的预算，所以它一半用玻璃制成，另一半用混凝土制成。最终，我们决定在上面铺盖假草坪。

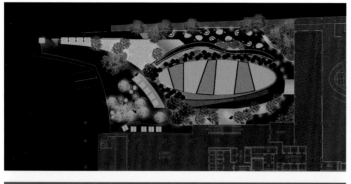

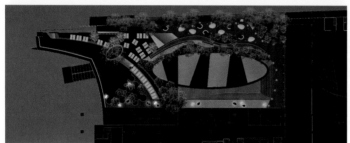

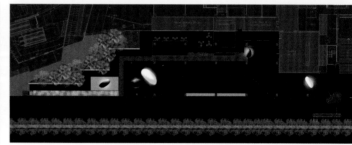

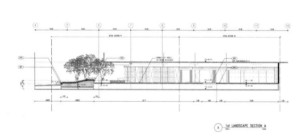

A 1st LANDSCAPE SECTION A

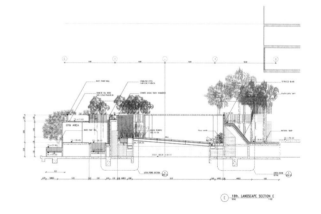

C 16th. LANDSCAPE SECTION E

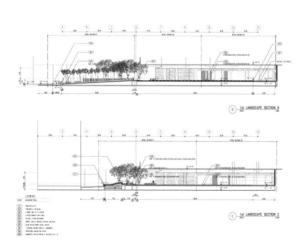

B 1st LANDSCAPE SECTION B

C 1st LANDSCAPE SECTION C

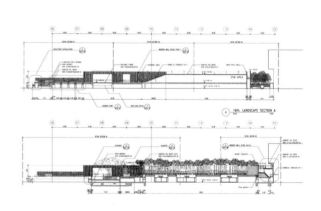

A 16th. LANDSCAPE SECTION A

2. The leftover areas on the roof.
Because the skylight is right in the middle of the roof, we ended up having only small and narrow areas around it for the garden.
Within our already-limited areas, we must locate gym and toilets somewhere as well.

第二，屋顶剩余区域。
由于天窗正好位于屋顶的中央，所以我们只能利用剩下的环绕在天窗周围的狭小区域建造花园。
在已经有限的区域内，我们还必须找地方安置体育馆和卫生间。

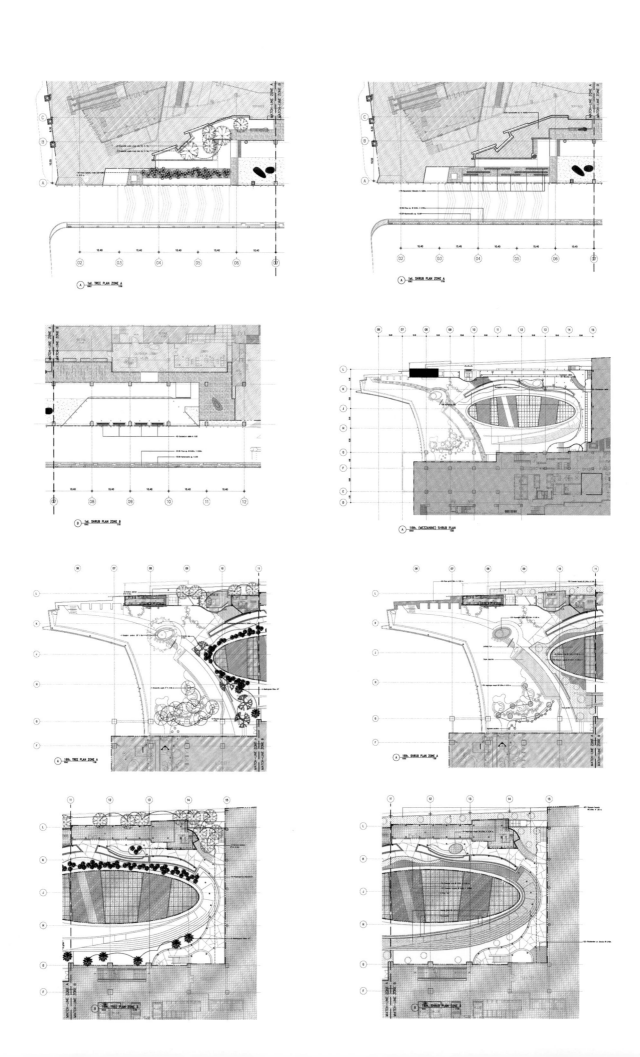

3. The irregular edge of the roof.
The mall has interesting internal and external facades. However, the edge made it so hard for us to design a clean and simple garden we want.
Our first move was to deal with the area around the skylight. We used the resort design principle to divide and reorganize the roof into three main garden parts.

第三，屋顶的不规则边缘。
购物中心拥有非常有趣的内外建筑表面。但是，屋顶的不规则边缘使得设计一个我们想要的整洁简单的花园变得异常艰难。
我们设计的第一步是处理天窗周围的区域。我们运用度假胜地的设计原理将屋顶分割重组成三大主花园区。

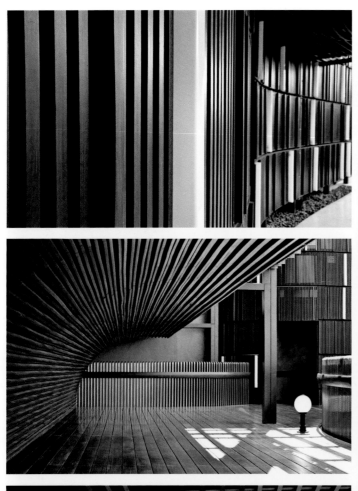

Noble Cube

来宝立方

LOCATION：Thailand
项目地点：泰国

AREA：51,222 m²
面积：51 222 平方米

COMPLETION DATE：2009
完成时间：2009 年

PHOTOGRAPHER：Pok Kobkongsanti
摄影师：Pok Kobkongsanti

DESIGN DIRECTOR：Pok Kobkongsanti
设计总监：Pok Kobkongsanti

TEAM：Pakawat Varaphakdi，Wasin Muneepeerakul，Teerayut Pruekpanasan，Kampon Prakobsajakul，Sorakom Klongvessa
团队：Pakawat Varaphakdi，Wasin Muneepeerakul，Teerayut Pruekpanasan，Kampon Prakobsajakul，Sorakom Klongvessa

DESIGN COMPANY：T.R.O.P
设计公司：T.R.O.P

Noble Cube

Noble Cube

LOCATION：Thailand
项目地点：泰国

AREA：51,222 m²
面积：51 222 平方米

COMPLETION DATE：2009
完成时间：2009 年

Client asked us to design their sales office first. The site is quite small and we have a very limited budget for the landscape. However, Noble Group is quite an interesting client. They really encourage design and love to try new things as long as they are good and fun.

Because of the budget, basically, we didn't have many choices for the materials. So we gathered a list of affordable materials and tried to find a way to make them more interesting.

最初，客户要求我们去设计他们的销售办事处。场地面积很小而且能用于景观建设的预算也很有限。但是，来宝集团是一个很有趣的客户。他们非常鼓励设计事业并且热爱尝试新事物，只要这个新事物优质并有趣。由于预算并不充足，我们在材料上并没有太多的选择。所以，我们列出了能支付得起的材料并努力想办法将它们运用得更加有趣。

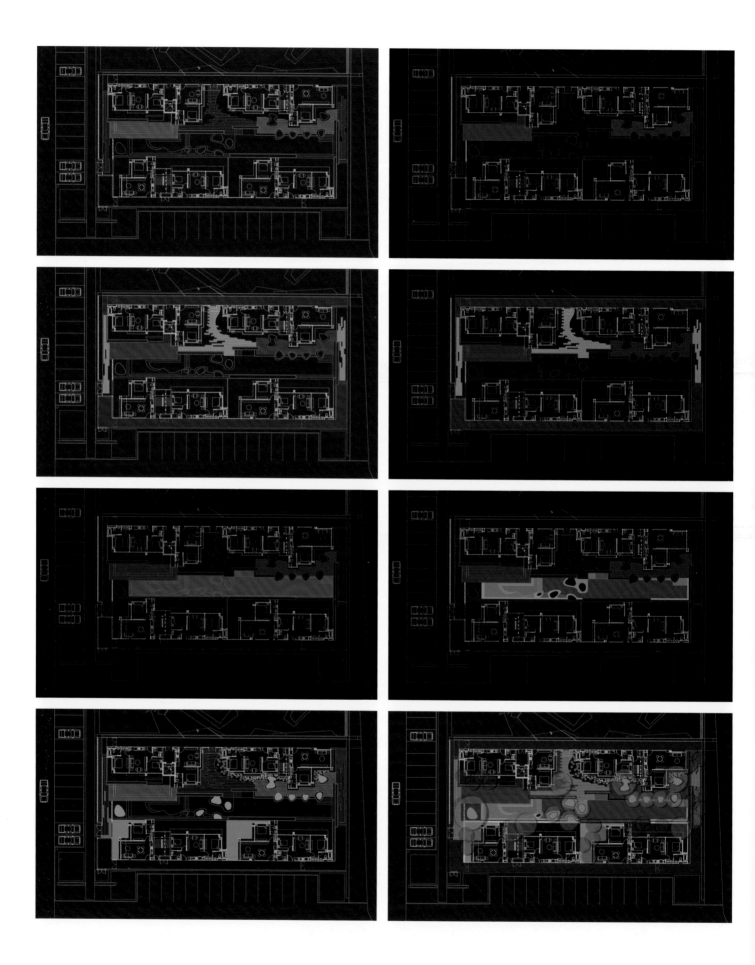

Our concept was inspired by paper collage paintings. Instead of colorful paper, we used different materials with different colors. We chose concrete, shipped rocks, lawn, some shrubs and other materials as our "colors".
We arranged the materials into a nice 2D composition first, which was like a painting, and then we started to transform the design into 3D by raising some pieces of materials here and there. Some of those become our main steps into the sales office, while others become garden elements.

我们的灵感来自纸质拼贴画。我们采用不同颜色的不同材料来代替彩色纸张。我们挑选了混凝土、舶来的岩石、草坪、灌木和其他材料作为我们的"颜色"。首先,我们将将材料整理成2D的平面结构,就像一幅绘画作品一样,然后,我们开始通过到处添加材料碎片将其转化成3D的立体图。完成的作品中,一部分成为通向销售办事处的主台阶,而剩下的用于修建花园。

Finally, we succeeded in building the garden, with a very small budget, as an interesting piece of green area. Because the location is right at the main traffic junction of the district, local people and the passers-by also love the project. Its green landscape helps them reduce stresses from the congested traffic around the area. And the customers of the project, mostly young professionals, love the distinguishing look of the garden.

最后，凭借仅有的少量预算，我们成功地建造了花园，并使其成为绿化带中很有趣的一部分。由于花园正好位于市区主要交通干道的交会处，当地居民和过往行人都十分喜欢这个项目。花园的绿色景观能帮助他们缓解由于周围拥挤的交通所造成的精神紧张。另外，这个项目的顾客大部分都是年轻的专业人士，他们很喜欢这个花园独一无二的景观。

Noble Remix

来宝混音

LOCATION：Thailand
项目地点：泰国

AREA：5,684 m²
面积：5 684 平方米

COMPLETION DATE：2011
完成时间：2011 年

PHOTOGRAPHER：Pok Kobkongsanti
摄影师：Pok Kobkongsanti

DESIGN DIRECTOR：Pok Kobkongsanti
设计总监：Pok Kobkongsanti

TEAM：Kampon Prakobsajakul
团队：Kampon Prakobsajakul

DESIGN COMPANY：T.R.O.P
设计公司：T.R.O.P

Noble Remix

What is the role of landscape architects in the world with global warming problem? How could we help reduce the heat wave? Could we encourage people to plant more trees? Those are some questions we have in mind all the time. And finally we have a chance to start adding some more "greenery" back to the urban landscape.

在全球变暖、温室效应频发的今天，景观设计师们应当扮演什么样的角色呢？我们怎样才能帮忙减少热浪？我们能否鼓励人们去多种树？这些是整日萦绕在我们脑中的问题。最后，我们有了一个向城市景观中注入更多绿色植物的机会。

Located at one of the best locations in Bangkok, Sukhumvit Road, Noble Remix is a residential project, with plus retails on its ground floor. The project targets at young professionals, who love modern way of living (with a little sense of humor). We got a commission to design its plaza on the ground floor. The area is sandwiched between the building and Sukhumvit Road. Our clients did not give us specific programs or tell us what they wanted. Basically, they just needed some green area to apply for the EIA permit, in order to use the building. So we have to find out what could and should be done there.

我们的任务——来宝混音，是一个住宅项目，底层用于附加的零售。该项目的选址位于曼谷的最佳地段——馨乐庭路。此次项目的目标受众群定位于那些喜欢现代化生活方式的年轻专业人士（幽默一点地说）。除此之外，我们还受委托在底层设计一个广场。广场的区域夹在楼房和馨乐庭路之间。我们的客户并没有给我们明确具体的计划书或是告诉我们他们的期许。基本上，他们只需要一些绿化带去申请获得环境影响评价许可，从而使人们可以使用这栋建筑。所以，我们必须去看看我们能做什么和应该做什么。

First, we studied the area of Sukhumvit Road. The road is not only the prime area for any project, bot also one of the worst traffic—jammed areas in Bangkok. The area would have a series of very bad traffic jams during a day, from early morning (6am—10am) to the evening (5pm—sometimes, midnight). The site, even though it is located on the great location, just has no great view on any side of the project. So, instead of looking out, we think that the plaza will be something that people can look at. It will be a new eye candy for passing by people (pedestrians, drivers, and train passengers).

首先，我们研究了馨乐庭路这片区域。馨乐庭路不仅是任何项目的最佳选址地，还是曼谷交通阻塞最严重的路段之一。这片区域从早上（上午6点到上午10点）到晚上（下午5点开始，有时持续到午夜）会有非常严重的交通堵塞情况。这个场地虽然位置非常好，但从楼层的任何一面都看不到好的景色。所以，我们认为，与其寄希望于外部，不如将广场打造成吸引人们眼球的景观。新建成的广场将会成为来往路人（包括行人、司机、火车乘客们等）新的视觉盛宴。

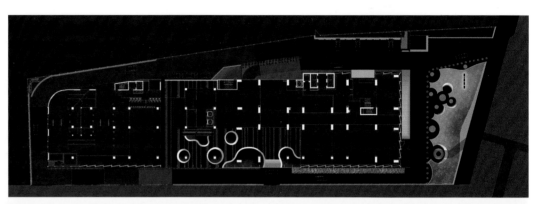

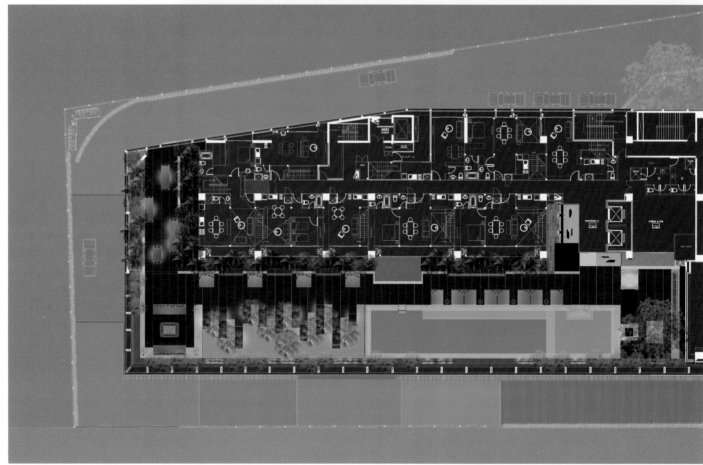

Naturally, the plaza can be viewed from two angles, from the building and from the road. The word "two-sides" became our first rolling point.

We thought about Thailand's old philosophy "a coin has two Faces", which can be applied to any human life. In one's life, a person could have two faces as well. On the front face, there are look, work, responsibility, taste, etc. On the backside, without being noticed, one may has lust, personality, demand of entertainment, etc.

自然而然地，广场可以从两个方面进行观赏，从楼层上观望或是在道路上欣赏。"两面性"这个词成为了我们此次项目的第一个焦点。我们思索了一则泰国古老的哲理——"每个硬币都有正反两面"，这句谚语可应用于任何人身上。对于人生来说，一个人同样可以拥有双面生活。在人生的正面，是外表、工作、责任、品味等等；在人生的背面，没有人感知的那一面，是欲望、个性、娱乐的需求等等。

Privé by Sansiri

Privé by Sansiri 公寓

LOCATION：Thailand
项目地点：泰国

AREA：1st floor =1,144 m², area14th floor =760 m²
面积：1 楼为 1 144 平方米，14 楼为 760 平方米

COMPLETION DATE：2010
完成时间：2010 年

PHOTOGRAPHER：Charkhrit Chartarsa, Pok Kobkongsanti
摄影师：Charkhrit Chartarsa, Pok Kobkongsanti

DESIGN DIRECTOR：Pok Kobkongsanti
设计总监：Pok Kobkongsanti

TEAM：Pakawat Varaphakdi，Anuwit Cheewarattanaporn，Naratip Bundi，Pattanee Ukam，Chatchawan Banjongsiri
团队：Pakawat Varaphakdi，Anuwit Cheewarattanaporn，Naratip Bundi，Pattanee Ukam，Chatchawan Banjongsiri

DESIGN COMPANY：T.R.O.P
设计公司：T.R.O.P

Privé by Sansiri

AREA：1st floor =1,144 m², area14th floor =760 m²
面积：1 楼为 1 144 平方米，14 楼为 760 平方米

COMPLETION DATE：2010
完成时间：2010 年

Privé by Sansiri is an exclusive luxury condominium in the prime Bangkok location. Our target group is successful people with age above 40, so the design has to be neat and elegant.

Privé by Sansiri 是位于曼谷核心地带的一座独一无二的高档奢华公寓。我们将目标消费群体定位于 40 岁以上的成功人士，所以公寓的设计务必整洁而且品位高雅。

T.R.O.P's scope included the ground floor garden and the swimming pool on the roof. For the ground floor, Sansiri asked us to create a wall to enclose and screen the lobby from the public. However, we found that the area is a bit small. So instead of building a solid wall, which would make the area feel even smaller, we proposed a custom-designed sculpture wall as the alternative. The wall is a series of sculptural columns, each has some space from one another. As a result, the area has some ventilation and plays with natural light in a much more interesting way. At the base of the columns, we strategically placed a reflecting pond to make the vision even more beautiful.

T.R.O.P 的任务包括底层花园和屋顶游泳池的设计和建设。关于公寓底层，Sansiri 要求我们围绕大厅修建一面墙，从而将大厅从公众视野中包围隐蔽起来。但是，我们发现空间有一些狭小。所以，如果真的建一面实体墙，那底层区域的面积则会显得更小。我们提议修建一面定制的雕刻墙代替实体墙。雕刻墙由一系列带有雕刻图案的圆柱体组成，圆柱与圆柱之间留有一定的空间。这样设计的效果是使底层区域通风良好并达到了自然光与影的和谐。在圆柱的底端，我们还策略性地设置了一个倒影水塘，增添视觉美感。

For the pool, originally, the architect proposed a small rectangular pool in the middle of the roof. Because we have a great view here, we suggested them to create an L—shaped pool, at the right edge of the building instead. With this design, we have one of the best views of Bangkok for our residents. Then we played with the composition of the pool terrace. We divided the terrace into several portions. In this way, a person would not see the whole garden at once. He has to walk around and discover some secret corners in the garden by himself. With a variety of space provided on the roof, everyone can use it without disturbing others.

对于游泳池，最初建筑师提议在屋顶的正中央修建一个小型的长方形泳池。但是，因为在屋顶我们能饱览极美的风光，于是，我们建议他们在紧靠公寓右边沿的地方修建一个 L 形的游泳池。通过这样的设计，我们便能为公寓的居住者们提供最棒的曼谷风光。然后，对于泳池阳台的设计，我们加了一点小技巧，将阳台分割成几部分。这样，居住者们就无法看到整座花园。他们必须到处逛逛，自己发现花园里的那些秘密角落。由于我们在屋顶分割出了多种多样的空间，每位居住者都可以使用屋顶而不会打扰别人。

Zense Gourmet Deck and Lounge Panorama

Zense 美食甲板和休息室全景

LOCATION：Thailand
项目地点：泰国

AREA：2,000 m²
面积：2 000 平方米

COMPLETION DATE：2009
完成时间：2009 年

LANDSCAPE ARCHITECT：Pok Kobkongsanti
景观设计师：Pok Kobkongsanti

DESIGN DIRECTOR：Pok Kobkongsanti
设计总监：Pok Kobkongsanti

TEAM：Wasin Muneepeerakul，Pakawat Varaphakdi，Kampon Prakobsajakul
团队：Wasin Muneepeerakul，Pakawat Varaphakdi，Kampon Prakobsajakul

DESIGN COMPANY：T.R.O.P
设计公司：T.R.O.P

Zense Gourmet Deck and Lounge Panorama

Zense Gourmet Deck is the first of the four new destinations, created by Zen World of Central Retails Corp. The main concept is the four elements, Earth, Water, Wind, and Fire. Zense's design inspiration comes from the Earth theme.

Zense 美食甲板是由 Zen 世界中央零售集团创建的四大新景区中的第一个。其设计围绕的主要概念来自四大元素——土、水、风和火。Zense 的设计灵感来自大地主题。

Zense 美食甲板是由 Zen 世界中央零售集团创建的四大新景区中的第一个。其设计围绕的主要概念来自四大元素——土、水、风和火。Zense 的设计灵感来自大地主题。

The location of the project is on the roof top of Zen World, which is on the 17th floor. Here, we have one of the best views of Bangkok's central area. The project started in 2007. The interior is designed by DEPARTMENT OF ARCHITECTURE, a famous young Thai architecture studio. T.R.O.P has been contacted by DEPT to help in landscape design of the exterior area. The main feature of this place is the unobstructed panoramic view. So our first design move is to make sure that we could even make the view nicer. The total exterior area is about 900 m² approx. The problem is that client wanted to add as many seats as possible. Also one major concern is that the existing shape of the deck is not rectangular, but rather half oval shape. So we started by creating interesting seating arrangements for each zone of the deck. In the end, we created some level differences in our main deck, in order to divide the space into several zones, like main dining area, private zone, and sky bar.

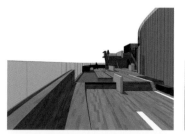
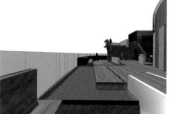

Zense 美食甲板位于 Zen 世界 17 层的顶层屋顶。那是观看曼谷城市中心区域风光的最佳地点之一。这个项目开始于 2007 年。其内部设计是由泰国一家著名的、年轻的建筑工作室——建筑部完成的。受 DEPT 之邀，T.R.O.P 公司协助完成外部区域的景观设计。项目所在地的最大特点便是可以观赏曼谷全景风光。所以，我们设计的第一步便是保证让风光看起来更漂亮些。外部区域的总面积大概有 900 平方米，但有一个问题，客户希望在外部区域尽可能多地加入座位；而且，最主要的顾虑是甲板现有的形状并不是长方形，而是半椭圆形。所以，我们工作的第一步便是为甲板的每一个区域创建有趣的座椅布置形式。最后，我们在主甲板区创建了不同的隔层，能将空间分割为几大不同的区域，比如主进餐区、私人区和天际酒吧。

As agreed from the beginning, the main material of the place must be wood, to represent the Earth theme. We selected the local wood plank, which is easy to find and can stand the extreme temperature of Bangkok's summer. The key is to pick a thicker piece, so it will not bend. T.R.O.P's goal is to create interesting space, using just one material, and also to make it mute yet urban. We want to create special experience for anyone who has a chance to visit the restaurant.

按照一开始的协议，甲板的主要建筑材料必须是木材，从而展现大地这一主题。我们选择了当地的木材厚板，因为它们很容易获取并且能承受曼谷夏日的极度高温。关键是选取厚一些的木段，因为厚一些的木段不容易弯曲。T.R.O.P 公司的目标是只用一种材料便可创建一个有趣的空间，营造一种既静谧又都市化的氛围。我们希望为有机会拜访餐厅的每一位顾客创造一份特别的体验。

Unlike most high—end restaurants, which normally use the most expensive plants available, the plants for Zense Gourmet Deck are basically the unwanted species. We use local plants, which normally grow at street sides, vacant lots, or dirty canals in Bangkok, in order to recreate the sense of place in this restaurant.

高端餐厅通常使用最昂贵的植物用于装饰，与这些高端餐厅所不同的是，Zense美食甲板所采用的植物基本上都属于不受欢迎的物种。我们选择了本土的植物，那些通常生长在曼谷的道路两边、城市空地，或是肮脏的河道里的植物。用这些植物来重置这家餐厅的景色。

However, we design, arrange and compose those plants in a more modern way. The result is quite interesting. People always know these plants, but never appreciate the beauty of them. They thought the plants are cheap and dirty. After they see our design, they have changed their perception of the plants. They start to see their beautiful look and cool feeling once again.

但是，我们以一种更现代的方式去设计、安置和组合这些植物。结果十分有趣。虽然人们一直很了解这些植物，但却从来不会去欣赏它们的美。他们认为这些植物廉价而且不洁净。但是，在看过我们的设计之后，他们改变了自己对这些植物的看法。他们开始重新审视它们，发现它们的美丽外表和清凉感。

surfacedesign, Inc

Awards, Competitions & publications
Garden Design Magazine, June 2011
California Home and Design, June 2011
Los Angeles Times, April 2011
Opening up Auckland's Waterfront, April 2011
San Francisco Chronicle, May 2011
San Francisco Chronicle, April 2011

New Zealand Institute of Landscape Architects Award of Excellence, 2010
American Society of Landscape Architects Honor Award, 2009
American Society of Landscape Architects Honor Award, 2007
Cofco Headquarters, Beijing, China, 2010 ——— Competition Finalist
BFS Gateway, Beijing, China, 2010 ——— Competition Finalist
Fan Hai International Center, 2009 ——— Competition Winner
Seoul Station Cultural Center, 2009 ——— Competition Winner

Planting: The Planting Design Book for the Twenty-First Century, Diarmuid Gavin & Terence Conran, 2009
California Home and Design Magazine "Ten to Watch", May / June 2009
Dwell Magazine "Beyond Green", May 2009
Landscape World vol 18, Korea, 2008
Best Private Plots Design Award, Topos Magazine, 2008
Landscape Architecture Magazine, January 2008
Environment and Landscape Architecture, Korea, October 2008
Small Firms Great Projects, 2008/2009
Gardens Illustrated, Issue 132, 2008
Landscape Design 31, China, November 2008
Landscape Design 30, China, September 2008
Landscape World vol 18, Korea, 2008
Environment and Landscape Architecture, Korea, October 2008
1000 X Landscape Architecture, Braun Publishing, 2008/2009
Die besten Gärten, 2008 ——— best private plots 08, Austria, September 2008
Landscape Architecture Magazine, January 2008
Gardens Illustrated, Issue 132, 2008
Architectural Record, January 2008
Garden Design, October/December, 2007
Dwell Magazine, June 2007
Urban Space Design, April 2007

James A. Lord

Education
Graduate School of Design in Harvard University, Master of Landscape Architecture, 1996
University of Southern California, Bachelor of Architecture with honors, 1990

Professional Experience
Surfacedesign, Inc.
Peter Walker and Partners
Martha Schwartz, Inc.
Hargreaves Associates
John Mutlow, Architects

Roderick R. Wyllie

Education
Graduate School of Design in Harvard University, Master of Landscape Architecture, 1998
University of California, Santa Cruz, Bachelor of Arts in Music, 1990

Professional Experience
Surfacedesign, Inc.
Lutsko Associates
Martha Fry Landscape Architects
Peter Walker and Partners
Martha Schwartz, Inc.

Geoff di Girolamo

Education
University of California at Los Angeles, Master of Architecture, 1993
University of California at Santa Cruz, Bachelor of Art, 1989

Professional Experience
Surfacedesign, Inc.
Marta Fry Landscape Architects
SOM(New York, San Francisco, Hong Kong)
SMWM
Pei Cobb Freed and Partners
New School University, Wagner School of Urban Policy and Management

Surfacedesign, Inc.

Surfacedesign, Inc.

Surfacedesign, Inc. was established in 2001 to provide clients with a broad range of professional design services, including landscape architecture, urban design, and master planning. The award-winning practice is engaged in projects of a variety of different scales, both locally and internationally: estate design, park design, hospitality, corporate campuses, municipal streetscapes, and large-scale land use planning and urban design projects. We create projects that have a strong relationship to people and the natural environment; we are passionate about craftsmanship and about creating sustainability. Under the leadership of James A. Lord, Roderick Wyllie and Geoff di Girolamo, Surfacedesign is an innovative and multidisciplinary design practice, employing the talents of landscape designers, planners and architectural designers. The artistry and durability of our work is the result of strong conceptual design, facility in working with materials and planting, and proven experience in project management and construction supervision. The firm's work methodology emphasizes understanding and accommodating the specific needs of each client and the unique programmatic and contextual requirements of each project. Our clients include corporations, real estate developers, architects, planners, public agencies, and homeowners.

Cow Hollow School Playground

Cow Hollow 学校操场

LOCATION：San Francisco, USA
项目地点：美国 旧金山

DESIGN COMPANY：Surfacedesign, Inc.
设计公司：Surfacedesign, Inc.

Cow Hollow School
Playground

During its short life as a public park and national landmark district, San Francisco's Presidio has become an enduring cultural and environmental resource for the City. Strict stewardship of the park is overseen by the Presidio Trust whose mission is to both protect and restore the resources of the park and to provide opportunities for residents, visitors and businessmen in the Presidio to access this unique place.

旧金山普雷西迪奥在它作为一个公共公园和国家地标区的短暂历史中，已经成为这座城市一个恒久的文化和环境源泉。普雷西迪奥信托公司负责对该公园进行严格的管理，保护和维护公园的资源，为当地居民、游客和商人提供享受这个无与伦比的人间乐园的机会。

When the Cow Hollow School, a preschool for children with 2—5 years of age, decided on a new home to expand its facilities and educational mission, the Presidio was a logical choice as the combination of nature and history coincided with the specific goals of its curriculum. Located a short walk from the Presidio's historic parade ground, the school contacted the landscape architect to conceive of a playground design that is an extension of the classroom. Extensive workshops with school teachers, administrators and parents helped the designers to understand the school's education mission: learning through play; exploration and discovery; and nurturing relationships between children, parents and teachers.

当 Cow Hollow 学校（一所招收 2-5 岁儿童的幼儿园）决定建一所新校舍，从而增添设施和教学功能时，普雷西迪奥由于将自然与历史相结合，符合了学校课程的特殊目的和要求，成为了一个符合逻辑的理性选择。学校位于具有历史意义的普雷西迪奥阅兵场附近，步行几分钟就可以到达阅兵场。学校向景观设计师表达了自己的想法：把操场作为教室的延伸。景观设计师与学校的教师、管理者和学生家长充分讨论，他们帮助景观设计师去理解学校教育的使命：在玩中学；探索与发现；培养学生、家长和老师的关系。

Curriculum at the school is based on the Reggio Emilia approach, a philosophy and educational program that teaches principles of respect, responsibility and community by encouraging children to be their own teachers and by providing a supportive and enriching environment based on the interests of the children. As the organization of the physical environment is linked to Reggio Emilia's early childhood program (the environment is often seen as the "third teacher"), the design of the playground is seen as critical to the curriculum of the school.

学校的课程是基于瑞吉欧·艾米利亚的教育方法，她提出了一种理念和教育方案：通过鼓励孩子自学的方式，以及根据孩子们的兴趣来提供丰富的辅助环境，让学生学会尊重、承担责任和交流。由于自然环境的安排状况与瑞吉欧·艾米利亚的学前教育计划（环境被看做是第三个老师）紧密联系，因此，操场的设计被认为是学校课程设计的关键。

The landscape design for the Cow Hollow School is indebted to nearby physical surroundings of the San Francisco Bay area and to the natural setting of the Presidio.

References to the local landscape typologies of beaches, forests, the Marin headlands and bay tidal wetlands are incorporated into the physical configuration of the school yard. The playground design incorporates the educational and interactive requirements of the Reggio Emilia program by emphasizing the landscape as an extension of the classroom, by providing a variety of exterior spaces for sel-instruction, and by emphasizing direct visual and physical connections between the classroom and natural landscape.

学校景观的设计得益于附近旧金山湾地区周围的自然环境和普雷西迪奥区的自然环境。

我们在布局学校庭院时参考了当地的沙滩、森林、马林陆岬和海湾潮汐湿地等典型的景观。该设计考虑了瑞吉欧·艾米利亚的教育方案中对教育和互动提出的要求，强调景观是教室的延伸，强调教室和自然景观之间直接的感观联系，为学生提供一个多样化的可自学的外部空间。

Getting building approvals to build the new playground from the Presidio Trust proved to be a formidable task for the client and the designers. The playground design needed to demonstrate that there would be no impact on the native plants habitat at the site. Historic and cultural resources below the surface of the site needed to be protected as well. All built and planted elements of the playground would have to be completely removable as to not disturb the ancient foundations of the original Spanish Presidio, the namesake of the park. As a result, trees and plant materials were placed away from the Presidio foundations and materials such as rammed earth and path fines were selected as to not impact antique below.

从普雷西迪奥信托公司得到的建设这个新操场的许可对于客户和设计者来说是一项艰巨而伟大的任务。操场的设计要体现出该项目对当地的植被不会有影响。该场址地下的历史和文化资源也需要保护。操场上建造和种植的所有东西必须是可完全移除的，为的是不破坏西班牙普雷西迪奥公园原始的古代地基。所以，树木植物被安置在远离普雷西迪奥古代地基的地方，建筑材料如夯土和路面细料都经过了精挑细选，为的是不影响下面的古文物。

Liberty Hill Residence

自由山住宅

LOCATION: San Francisco, USA
项目地点：美国 旧金山

DESIGN COMPANY: Surfacedesign, Inc.
设计公司：Surfacedesign, Inc.

Liberty Hill Residence

Marked by its dramatic sloping topography, this private garden in San Francisco's Liberty Hill has dedicated areas for entertaining and children's play, which are defined and navigated by an innovative arrangement of hardscape materials. Cor—Ten steel boxes serve as retaining structures and planters, extending along the site's perimeter and penetrating the surrounding wood fence. Taken as a whole, the back yard is an abstract allusion to the hilly city itself, which appears to fall away behind the garden when viewed from the upper terrace or out the back windows of the house.

该地区的特征是斜坡地貌，这所旧金山自由山上的私人花园有专门的娱乐场所和儿童游乐场地，该花园的特点是具有创新性的人造景观素材。耐腐蚀高强度钢箱充当固定结构和花盆，沿着该住所的周围延伸，然后深入周围的木栅栏。从总体上看，后院暗示了这座山城的存在，当从这所房子的后窗或从高处的露台上眺望，城市看起来在花园后面渐渐消失了。

The Cor—Ten steel boxes and concrete walls are added to existing stone retaining walls to create a series of switchbacks that guide the journey from the house into the garden. Erected in mid—century, the stone walls form a natural part of the vocabulary of rugged and refined materials —stone, concrete, wood, and steel—while uniting the site's past and present incarnations in a visually harmonious way. The steel boxes, which harbor plant life up top, descend into the ground plane and insert into the redwood fence—a geometric composition enhanced by the slanting shadows that are cast through the variegated slats. The dimensions of the slats are echoed in the board form pattern on the concrete walls that are wrapped in the fence.

耐腐蚀高强度钢箱和混凝土墙与现有的石墙创造了一系列的Z字形路线，这条路会把你从房屋带入花园。中世纪的石墙是天然的材料，属于精致坚固的建筑材料中的一部分，这些材料有石块、混凝土、木头和钢结构，它们以一种可见的和谐的方式体现这个住所的过去和现在。这个钢结构的金属箱在上部为植物提供了庇护所，向下与地平面相接，深入红木栅栏，斜影穿过色彩斑斓的板条，增强了栅栏的几何图案。板条形状的阴影在混凝土墙上显现出来，仿佛墙被栅栏包裹起来。

Seeking to maximize permeability for the overall health of the garden, the large children's area is lawn and the entertainment terrace is decomposed granite. The planter boxes not only afford the homeowners an opportunity to do some hands—on gardening, they also attract pollinators such as birds and butterflies. The concrete steps are framed by runnels to enhance drainage (while highlighting the sloped nature of the garden), and a native sedge called Juncus helps purify run—off before it enters the groundwater system. Because the steel boxes are framed by the runnels, they appear almost to rise from below the ground plane, adding depth to the composition.

为了最大化地渗透花园的健康主题，孩子们的大活动区域是草坪，风化花岗岩筑成娱乐的平台。一些大的花盆不仅为主人提供亲自动手从事园艺的机会，还可以吸引小鸟和蝴蝶驻足栖息。混凝土台阶的两侧都是水流，可以方便排水（同时突出花园倾斜的特点），一种叫做灯心草的当地植物起到了在污水进入地下水系统之前净化水流的作用。因为钢制箱子的边上也有水流，它们看起来好像是从地面下升起的，给整个构图增添纵深效果。

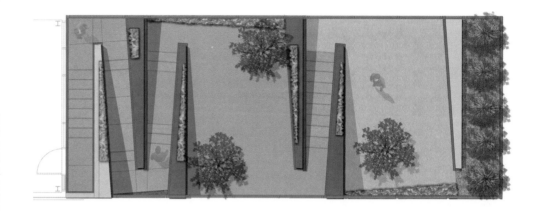

As the homeowners spend much time looking down the garden from their homes, they also appreciate its sculptural quality, which is punctuated by plant materials such as Japanese maples and the softening presence of shade—lovers such as ferns, irises, viburnum and anemones.

当房屋的主人从屋里长时间向下观看花园的时候，他们也会欣赏到雕塑上的美，它们点缀着诸如鸡爪枫以及蕨菜、鸢尾花、荚莲、银莲花等喜阴轻柔的植物。

Peterson Residence

Peterson 居所

LOCATION: San Francisco, USA
项目地点：美国 旧金山

DESIGN COMPANY: Surfacedesign, Inc.
设计公司：Surfacedesign, Inc.

Peterson Residence

Situated on a peninsula overlooking San Francisco Bay, this sculptural garden reflects the unique vision of passionate art collectors whose personal artistic impulse includes the idea of the landscape being experienced and enjoyed by their children. As a result, the garden incorporates several distinct spaces that weave together architecture, landscape, and use, while seeking to capture the site's unique views and topography.

这个雕像主题公园位于一个能够俯瞰旧金山湾的半岛上，反映出了充满激情的艺术收藏家们独特的想象力，他们个人的艺术创作冲动部分是源于自己孩子对景观的感受和喜爱。因此，这座花园的设计在努力表现该地独特的景色和地形的同时，还融入了几个迥然不同的空间，这些空间把建筑、景观、功能融为一体。

At the site's entrance, a folding deck projects from the interior of the residence, merging the house with a landscape of magnolias, grass, and ferns. The deck's warping planes create a threshold for the visitor, highlighting the relationship between the existing ground plane of the house and the sudden change of the surrounding topography. Beyond the deck, an existing cobbled pathway connects with a newly-planted shade garden, creating a sense of calm and intimacy around the main entrance. Opposite the main entrance path, locally-sourced river stones artfully conceal a main drainage swale. Tying into the existing slope. Past the front door, lush groves of bamboo and shade-loving trees frame the lyrical movement of a meandering path that terminates in a sculptural focal point surrounded by hillside grass. The solidity of this hillside sculpture juxtaposes with generous sweeps of grass, referencing the native wind blown landscape. The approach opens up to the east lawn area, offering extensive views of San Francisco Bay and the rolling California hills to the north.

在花园的入口处，从住宅内部伸出的双层露天平台将房子与外面的木兰花、草坪和蕨类植物美景连接在一起。露天平台下面的弯曲空间为游览者创设了一个入口，凸显了现有地面平台与突然改变的地貌之间的关系。越过平台，一条由鹅卵石堆砌而成的小径与林荫满地的花园相连接，给主入口营造出安逸、温馨的气氛。在主入口的对面，当地小河中的石块间巧妙地藏着一个主排水渠，水流流进斜坡。走过前门，郁郁葱葱的竹林和喜阴的树木造就了曲径通幽的林间小路，这条小路的尽头是一个引人侧目的雕像，由山坡的草环绕着。坚固的雕塑旁侧是开阔的草地，让人联想起当地风吹田野的景象。有路通到东侧的草坪，可以欣赏旧金山湾广阔的风景和北侧连绵的加利福尼亚山脉。

Reused granite curbing leads the way to the lower lawn where a succession of green mounds abstracts the distant hilly peninsulas that penetrate the Bay beyond. To the north of the house, a sequence of evergreen hedges and fragrant perennials create a serene viewing garden, aligning with the interior architecture.
The green hedging serves as an abstractdatum highlighting the twisting movements of a kinetic sculpture by George Rickey.

旧的花岗岩路沿石块铺成的小路通到下方的草坪，在那儿一连串的草丘让人想到远处深入海湾丘陵状的半岛。在房子的北面，一排排的常青树篱和芬芳的多年生植物构造出一个宁静的观赏园，与内部的建筑和谐地排布着。树篱起到了背景烘托作用，突出地表现出 George Rickey 大师的动态雕塑流动的曲线。

Children's play areas abound throughout the site. The folding deck at the site's entrance doubles as a climbing wall for the homeowners' active kids. The focal point of the children's area is defined by the playfully sculpted green mounds affording a wonderland of discovery and distinct play spaces; "Rabbit Hole Hill" allows children to drop balls into tubes of various sizes, without knowing where they will emerge in the landscape; "Jump Mound" conceals a hidden trampoline; "The Rabbit House", situated beyond a play structure, gives the children a personal play space among a landscape of native grasses. Similarly, the re-used granite curbing and curved lawn embankment are frequently used as an impromptu children's theater for open-air performances.
The overall sculptural quality of the landscape weaves together the homeowners' passion for their children and art collection by providing places for reflection, play, and discovery, while acknowledging the native, regional landscape.

这块场地里有许多供孩子娱乐的区域。入口处折曲的露台还可以成为房子主人活泼的孩子攀爬的墙。那些雕琢得可爱的绿色丘包是孩子们玩耍区域的焦点，提供了一个发现探索的奇境和不同的玩乐空间："兔穴山"可以让孩子们把球扔进不同尺寸的管子里面，不知道它们会从景观的哪里出来；"蹦跳土丘"遮蔽了一个隐蔽的蹦蹦床；"兔子房"位于一座娱乐设施的远端，在一块由本地草类组成的景观中给孩子们提供了一个私人娱乐空间。同样，重新利用起的花岗岩所砌成的路缘和曲线的草坪筑堤经常被当作孩子们露天即兴表演的舞台。整个景观的雕塑性特质与主人对孩子和艺术收藏的热情编织在一起，在体现当地区域风景的同时，提供思考、娱乐和探索的空间。

Rieders Residence

Rieders R 居所

LOCATION：San Francisco，USA
项目地点：美国 旧金山

DESIGN COMPANY：Surfacedesign，Inc.
设计公司：Surfacedesign，Inc.

Rieders Residence

The hill embarks fifteen feet from the house, which feeds into the garden on two levels. Entering through the garage from the street, the first view is a retaining wall, which is in dire need of replacement. On the main living level, the master bedroom opens onto the primary expanse of usable flat space. The challenge (and opportunity) is to create a cohesive plan for the vertical space, creating an experience of the garden below, providing a hospitable gathering space adjacent to the house, and providing access to the top of the site, with its postcard views of the ocean.

小山距离住宅 15 英尺，房屋从两个层面融入花园。从马路穿过车库进入住宅，首先映入眼帘的是挡土墙（护墙），但这一结构急需改变。在主要生活空间，主卧室朝向可用的平坦广阔空间。这个设计面临的挑战（也是机会）就是为这一垂直空间构思一个统一的计划，创造一种能感受到下面花园的体验，提供一个紧邻住所的会客空间，还要提供一个可以进入这所房子顶端的入口，因为从这里可以看到美丽的海景。

On the lower level, replacing the crumbling retaining wall offers an opportunity to incorporate the water feature requested by the homeowners. A fountain is embedded into two tiered walls; as water passes through bronze weirs, it cascades down into pools, creating a pleasant aural backdrop throughout the garden. Visually integrated into the natural environment, the board-form concrete used for the walls has a wood-grain pattern that lends textural integrity, providing erosion control without the weight and monotony of a monolithic slab. At its base, the wall is massive enough to plant in, replacing a once desolate space with a kind of tranquil green grotto and contemplative place to pause.

在低层，将破旧的挡土墙替换掉，这样就可以有机会加入房子主人要求的水景。一眼泉水从高低错落的两道墙中间穿过。当泉水流过铜质的堤坝时，就会像瀑布似地落在水池中，潺潺的流水声萦绕在花园中，让人陶醉其中。砌墙所用的混凝土板有木纹图案，这可以在视觉上融入自然环境，有了统一的质地，这样，无须使用巨大的石板就可以防止侵蚀，以免石板所带来的单调和沉重。墙的底部十分巨大，足够在墙体内栽种绿植，这样的设计把曾经一度荒凉的地方变成了一个绿色的静谧的石室空间，主人可以在这里停下来沉思。

Steps lead up to the main level, an expansive, semi—sheltered outdoor room ideal for dining and a safe play area for the children. Here the goal is to retain an existing oak tree and create harmonious plantings reflective of a natural Northern California plant palette, with native grasses and other drought—tolerant plants surrounding the oak with a sea of muted greens and dusky plums that harmonize against the backdrop of stained black cedar decking and walls.

顺着楼梯就可以走到主楼层，这是一个宽敞的、半遮蔽的户外理想就餐场所，也是孩子们安全的娱乐场所。我们的目标是保留现有的橡树，创造一个和谐、温馨的园林，这个园林能够表现北加利福尼亚植物缤纷的色彩，另外，当地的草类和一些耐旱的植物围绕在橡树周围，夹杂着许多色彩柔和的绿色植被和李子树，所有这些都以带有黑色斑纹的雪松木制甲板和墙体为背景，使得整个景色和谐美丽。

From here, the designers continue the circuit with in the grotto space below and with stairs gracefully leading up to a viewing platform. Previously inaccessible due to the steep grading, this panoramic aerie offers sweeping views of sunsets over the Pacific Ocean and landmarks such as the Golden Gate Bridge, providing a sense of place within the neighborhood, the city, and beyond.

从这里，设计师设计了一个螺旋上升的环形道，环形道的下面是石室，可以沿着环形道上典雅的环形楼梯走到观景台上。在这之前，由于过于陡峭，人们无法进入观景台，现在，人们可以步入这个高高的平台之上，抬眼望去，太平洋上日落之美景、像金门桥这样的地标性建筑尽收眼底，在这里你会觉得你与这个地区、这所城市以及所有的地方都融为一体。

Planting themes in the garden vary by location. The quiet lower grotto is composed of geometric concrete paving, metal planters, and lush, shade—loving plants, such as Woodwardia radicans, Helleborus orientalis, and Persicaria Red Dragon. The sunny hillside is covered with a crimson sea of Tradescantia pallida accented with a mix of agaves (Agave attenuate, Agave parryi, and Agave Blue Glow), Aeoniums and Sedums. The edges of the dramatic hillside and board—formed walls are softened with wispy plumes of Pennisetum alopecuroides. A small path of square concrete pavers leading to the top deck appears to float above a soft bed of nave fescues and is bordered by a mix of Deer Grass (Muhlenbergia rigens) and Kangaroo Paw (Angiozanthus flavidus).

花园绿化的主题是因地制宜的。安静低矮的石室由几何形的混凝土块石铺砌而成，另外这里还有金属花架和繁茂的喜阴植物，如狗脊蕨、圣诞蔷薇、红脚鹪。在向阳的山坡上覆盖着大片的紫叶鸭跖草，被各种龙舌兰、Aeoniums 花和景天属植物衬托着，引人注目的山坡和石板墙的边缘长着一束束纤细绒绒的狗尾草。沿着方形石板砌成的小路走到顶层平台，你会觉得自己仿佛漂浮在一个用嫩牛毛草做成的、鹿草和袋鼠爪做边的柔软舒适的大床上。

Bjørbekk & Lindheim

Featured large-scale projects

• Tjuvholmen. A new city district on the waterfront in Oslo next to Aker Brygge. In 2012, the entire district will be completed. Tjuvholmen will have high-end shops, offices, apartments, galleries, parks and a beach.
• Nansenparken, Fornebu. A park in the centre of the new large-scale housing area at Fornebu — this was formerly the main airport site of Oslo. The park opened in August 2008
• Pilestredet Park. An old national hospital being transformed into an attractive recreational and residential area.
• The Norwegian Radium Hospital. Outdoor park and traffic areas in connection with extension of the hospital
• The new central hospital for Akershus. The project includes planning of an outdoor park and traffic areas. This is the largest ongoing building project in the country at the present time. First phase of the project opened in 2008 and the construction of buildings, parkland and roads will continue until 2012.
• Hundsund grendesenter, Fornebu. Village center—outdoor areas for school, kindergarten, athlete hall, swimming hall and large outdoor ballgame arenas—p opened autumn 2008
• Fornebu: terrain and landscape management project at Fornebu—a former airport area.
• State road Rv 150 – Ring 3 – Ulven-Sinsen, Oslo: Projecting of new streets and roads—currently under construction.
• Ensjø new development area in Oslo for public housing: Project includes public squares, streets and parks, waste water management, opening of a stream.
• Bjørvika, Oslo. Detailing of streets in a new developing part of town.

Tone Lindheim and Jostein Bjørbekk are the founders and owners of Bjørbekk & Lindheim Landscape Architects. They are also project directors in the firm.

Tone Lindheim
She started her career as a project manager for urban renewal projects in Oslo from 1981—1984, and then worked as a landscape architect for Asplan Project in Oslo from 1984 until 1986.
As a practicing landscape architect, she has in particular directed large urban transformation projects (former airport, hospital, industrial sites) and urban parks, urban spaces, housing areas and schools. Some important and much publicized recent projects have been the prize—winning Nansen Park and Pilestredet Park housing area, as well as plans for the new urban area at Ensjø and Ekeberg Sculpture Park.
Since 1996, Lindheim has been a professor at the Institute of Landscape Architecture and Spatial Planning at the Norwegian University of Life Sciences in Ås.
Lindheim has been jury chairman and a member of juries for many competitions. She lectures extensively in Norway and abroad. She has wide editorial experience, and is the author of several books; *The Good Back Yard*, 1987; *The New Oslo*, 1987; *Noise Baffles in Oslo*, 1995 and *School Yards in Oslo*, 1999.

Jostein Bjørbekk
He started his career in landscape architecture when he worked as the project manager for landscaping of new headquarters of The State Bank of Norway in Oslo from 1982—1986, in the firm Hindhamar, Sundt, Thomassen.
He has been project manager for a number of major landscaping projects such as the new Akershus University Hospital (2000—2012), the Tjuvholmen retail and housing district outside Aker Brygge in the Oslo harbour, municipal streets in Bjørvika, a new district of Oslo's inner harbor, for the new highway and Storøya Recreation Area transformed from a former airport at Fornebu, outside Oslo. In the last three years he has also directed projects with further development of towns and villages in several municipalities in the northwestern parts of Norway.
He is currently working on the project Bogstadveien—renewal of Oslo's main shopping street, and the pilot projects for the landside at Flesland Airport, Bergen and the landside at Oslo Airport, Gardermoen.
Bjørbekk has been head of the Norwegian Landscape Architects Competition Committee from 2002—2009 and a member of the Architecture Council in Oslo municipality from 2002—2011.
In 1997 he was awarded the Norwegian Public Roads Department's Award for Beautiful Roads for the project Mannheller ferry quay.

Bjørbekk & Lindheim

A presentation of the studio

Jostein Bjørbekk and Tone Lindheim founded Bjørbekk & Lindheim Landscape Architects in 1986. The firm today is one of the leading Nordic firms within landscape architecture.

In our projects we listen very carefully to the site. Valuable elements on the site often give clear design guidelines in the process and the final concept for the project. We wish to honestly reflect the present time and therefore give our projects a clear contemporary contribution.

It is not only important to look for finds and show them, but also to add something specific and strong from our own time.

We try to create beauty in everyday life, to make attractive meeting places and functional landscapes which will age with dignity and maturity gracefully.

Our firm focuses on ecological solutions. We are involved in huge transformation projects; turning the old national airport, Fornebu, into a beautiful area with parks, recreational and residential areas. In another project, Pilestredet Park, we have turned the old national hospital into housing, parks and recreational areas in the centre of Oslo. We also work with old harbours and industrial sites.

Bjørbekk & Lindheim has received several 1st prizes, awards and nominations.

AHUS Hospital

阿克什胡斯大学医院

LOCATION: Akershus, Norway
项目地点：挪威 阿克什锦胡斯

AREA: 145,000 m²
面积：145 000 平方米

COMPLETION DATE: 2011
完成时间：2011 年

PHOTOGRAPHER: Bjørbekk & Lindheim AS
摄影师：Bjørbekk & Lindheim AS

DESIGNER: Bjørbekk & Lindheim AS /Schønherr Landskab
设计：Bjørbekk & Lindheim AS /Schønherr Landskab

LANDSCAPE ARCHITECTS: Bjørbekk & Lindheim AS and Schønherr Landskab with Jostein Bjørbekk and Torben Schønherr
景观设计：Bjørbekk & Lindheim AS，Schønherr Landskab，Jostein Bjørbekk ，Torben Schønherr

AHUS Hospital

The Oslo Landscape Architects Bjørbekk & Lindheim and Schönherr Landscape from Aarhus in Denmark have been responsible for the landscaping of Akershus University Hospital, known as AHUS, and located at Lørenskog which is a neighbouring municipality to Oslo. The old hospital on the site has been partially demolished and the rest will either be renovated or replaced by the new 140 000 m² (1,506,950 ft²) complex.

阿克什胡斯大学医院的景观设计由来自奥斯陆的景观建筑公司 Bjørbekk & Lindheim 和来自丹麦奥尔胡斯的 Schönherr Landscape 负责。阿克什胡斯大学又被称为 AHUS，位于与奥斯陆毗邻的城市 Lørenskog。原来的医院一部分已经拆毁，剩下的部分将进行翻新，或建造 140 000 平方米（1 506 950 平方英尺）的新综合楼。

The landscape surrounding the hospital would be completed during the final phase of the building works in 2012. In 2010, the forecourt and park outside the last completed inpatient building was under development.

医院周围的景观计划在 2012 年建筑工程的最后一期完成。在 2010 年，最后建造的住院部大楼外面的前院和公园正在开发之中。

The landscape surrounding the hospital
医院周围的景观

The landscape in the area is a forested hillside interspersed by open fields with streams and small rivers. The hospital complex is in the crest of a ridge surrounded by a 14 hm² (34.5 acres) park.

景观所在地是一个草木丛生的山坡，其间点缀着空地和小溪。医院综合体大楼位于一个山脊的顶部，周围被一个面积为14公顷（34.5英亩）的公园所环绕。

The goal of the landscape architects is to create the park as a refinement of the existing natural and cultural landscape that surrounds the area. The parkland itself is designed to allow accessibility to paths, open spaces and seating areas for hospital employees, patients, guests and the general public.

景观建筑师旨在通过提炼场地周围现有的自然和文化景观对公园进行打造。公共绿地本身的设计要保证医院的员工、病人、客人以及大众能够使用公园中的道路、空地和座位区。

The main elements of the hospital grounds are the forest, the grassy parkland and the forecourt.

森林、公共绿地和前院是医院景观的主要元素。

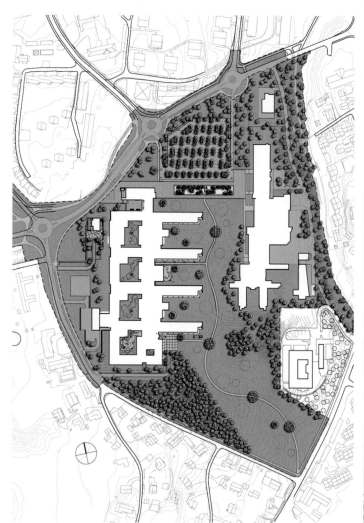

Forest
森林

The area where the hospital has been built is surrounded by areas of dense clusters of local birch, pine, willow and aspen trees to the east and west. This type of open forest landscape is strengthened by planting local trees on the hospital grounds.

医院所在的场地被东侧和西侧茂密的本地桦树、松树、柳树和白杨树丛环绕。医院所在的地块上也种植了一些本地树木，进一步加强了这种开放的森林景观。

At the parking area to the north of the forecourt, a row of trees will be planted that will, when fully grown, form a dense leafy cover, creating a "forest" car park.

前院北侧的停车区将会种植一排树木，树木长成以后会形成茂密的树荫，打造出一个"森林"停车场。

The green parkland
公共绿地

An important element of the hospital grounds will be the open grassy range which will stretch from north to south winding between the large, new buildings and the modernised parts of the old hospital. The parkland itself is characterised by small hilly mounds with dense circles of trees, reminiscent of the clusters of trees in the fields of the local landscape. Several of the new clusters are planted along the paths with ground lighting directed up at the tree crowns, providing unusual evening footpath lighting.

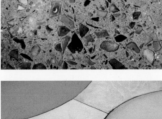
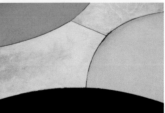

医院的一个重要元素就是在新建筑与原医院现代化区域之间打造从北向南蜿蜒延伸的开敞绿地。一个个小山丘形成公共绿地的特色，山丘上生长着茂密的树林，使人联想到了当地景观中的树丛。小路旁边种植一些新的树丛，地面上的照明直接照向树冠，在小路旁打造出一派不同反响的夜间照明景象。

One of the park's curiosities is the unbroken solid bedrock that forms the flooring of the public café in the glass—walled street. The rocky top of a hill has been wire—sawed and then sanded to reveal the composition of the stone. This is real bedrock, not stone slab flooring.

公园最引人注意的莫过于玻璃幕墙街道上构成公共咖啡屋地面的完整基岩。山丘的岩石表面用钢丝锯进行切割，然后铺上砂子，露出石头的组成形状。这是真正的基岩，而不是石头地面。

The green range forms the spine of the landscaping project and ends in the north at the large forecourt in front of the main entrance.

绿地形成了景观项目的脊柱，在主入口前面的大院北部截止。

Hundsund Community Centre

Hundsund 社区中心

LOCATION: Wiki, Norway
项目地点：挪威 维基

AREA: 50,000 m²
面积：50 000 平方米

COMPLETION DATE: 2009
完成时间：2009 年

AWARD: The National Building Design Prize in 2009
奖项： 2009 年国家建筑设计奖提名

ARCHITECTS: Div'A arkitekter
建筑设计：Div'A arkitekter

LANDSCAPE ARCHITECTS: Bjørbekk & Lindheim
景观设计：Bjørbekk & Lindheim

Hundsund Community Centre

Hundsund centre is built around a central pedestrian zone that is designed to pull together the fabric of the centre and provide access to all the facilities. The new buildings for the junior high school, the nursery school, the sports grounds and the indoors swimming pool have their own entrances from the precinct. Vehicles approach the precinct for loading and unloading at a roundabout at the beginning of the central zone.

Hundsund 中心建在一个中央人行区周围，其设计意在将中心的建筑组织在一起，使人们方便利用所有设施。初中、托儿所、运动场和室内游泳池等新建筑都有各自的入口。装卸的车辆可以从中心区起点处的环形道进入。

The area is designed to become a social meeting point for the new residents of Fornebu. The granite flooring of the central zone has water running in a curved gulley through it as a refreshing element.

这个地区将成为 Fornebu 新居民的社交集会地点。一条弯曲的水沟贯穿整个中心区的花岗岩地面，水在水沟中潺潺流淌，成为空间中的亮点。

The school and nursery school to the east and west of the precinct each have their own spacious outdoors areas. The buildings are surrounded by partly covered patios made of larch heartwood that open out to the south in large terraces with long tables and benches. Here children can participate in outdoor lessons and work on projects. The junior high school has a double level outdoor activity area. On the lower level, there are a series of structures and a stage for performances, as well as roped units to encourage climbing, outdoors meeting places on the upper level. There are also three hollow half-circles made of poured concrete for skating, biking, running and sliding. The outdoor area of the nursery school has a winding bike path moulded with ups and downs, a wooden pier, a water canal with a pump, a ladybug inspired knoll covered in soft-fall safety grass, a village of five small wooden playhouses. There are also a little kitchen garden, willow shrubbery that provides hidey-holes and a "Hundred Acre Wood" to play in.

东侧和西侧的学校和托儿所都有宽敞的室外区域。建筑被一部分带顶的中庭所围绕，中庭用落叶松心材建成，朝南开放，有巨大的台阶和长长的桌椅，在这里，孩子们可以参加室外课程和项目。初中有个两层的室外活动区，下层有一系列结构、一个表演舞台，上层有鼓励孩子们进行攀岩的带绳索的设施，以及室外聚集区。这里还有三个用浇筑混凝土建成的空心半圆，可以用于溜冰、骑自行车、奔跑和滑行。托儿所的室外空间有一条模制的蜿蜒起伏的自行车道，一个桩墩，一条带有坡道的水道，一个类似七星瓢虫的圆丘（上面用坡度较缓的安全玻璃覆盖），和一个由五个小屋组成的村庄。这里还有一个小菜园和一个柳树林，柳树林里面有一些可以用于捉迷藏的洞和一个"百英亩树林"，孩子可以在里面玩耍。

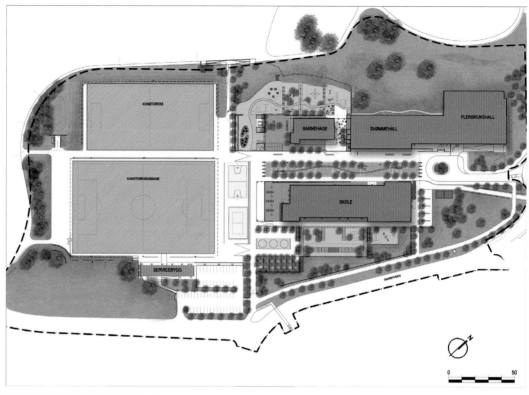

There are astro-turf playing fields, an ice skating rink and basket and beach volleyball courts to the south of the precinct.

社区中心的南部还设置了用人造草皮铺成的运动场、溜冰场、篮球场和沙滩排球场。

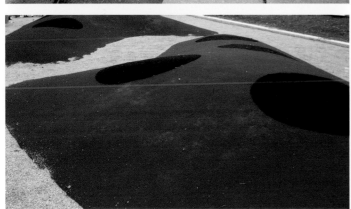

The Nansen Park, Oslo

南森公园，奥斯陆

LOCATION：Oslo，Norway
项目地点：挪威 奥斯陆

AREA：200,000 m²
面积：200 000 平方米

COMPLETION DATE：2008
完成时间：2008 年

LANDSCAPE ARCHITECT：Bjørbekk & Lindheim
景观设计：Bjørbekk & Lindheim

The Nansen Park, Oslo

An old cultivated landscape with much variation and beauty was levelled into Oslo International Airport in the 1940— 1960's. In 1998, the airport was moved and left behind it a peninsula of almost 1,000 acres in need of transformation.

20 世纪 40 年代到 60 年代期间，为了建造奥斯陆国际机场，这里变化多样的秀丽培植景观被夷为平地。1998 年机场搬走以后，留下一个近 1 000 英亩的半岛急需改造。

The moving of the Oslo International Airport at Fornebu resulted in the largest industrial reclamation project in the country. The new park was to form a functional focus and an identifying centrepiece of a new community some 10 km from downtown Oslo. Plots for housing and offices were sold off to private developers, while the Norwegian Directorate of Public Construction and Property together with the City of Oslo undertook responsibility for infrastructure and landscape, the treatment of polluted grounds, and the planning of a new park structure.

Fornebu 奥斯陆国际机场的搬迁，使得这里的改造成为国家最大的工业垦殖项目。新建公园主要以功能为主，形成一个地标，距离奥斯陆市区约 10 公里。用于住宅和办公的地块卖给了个人开发商，而基础设施和景观的建设，污染土地的处理以及新公园结构的规划则由挪威公共建筑和财产管理局和奥斯陆市负责。

In 2004, an architectural competition was won by landscape architects Bjørbekk & Lindheim.

2004 年，景观建筑师事务所 Bjørbekk & Lindheim 在本项目的建筑竞标中中标。

The central Nansen Park (approx. 200,000 m²), has been designed to serve as an attractive and active meeting place for all those who will be living at Fornebu. A strong identity, simplicity and timelessness have been key points. In order to respond to its dramatic history, the park has been designed as a dynamic dialogue between the uncompromising linearity of the airport and the softer, more organic forms of the original landscape. The site borders the Oslo Fjord on three sides. The openness of the landscape, as well as the distant contours of the hills gives a strong and peaceful feel of the sky, a separateness and spaciousness which we have tried to instil in the new landscape. The quiet calm of the extensive views and the harmonious forms have carefully been combined with activities.

南森公园中心（约 200 000 平方米）被设计成一个充满活力并且具有吸引力的场所，供 Fornebu 市民聚会之用，主要强调特性、简约和永恒。为了与其生动的历史相呼应，公园在机场生硬的线性和原有景观柔和、有机的构造之间形成生动的对话。这里三面都与奥斯陆峡湾相邻，景观的开敞性和远处小山的轮廓给天空带来一种强烈的平和感，一种我们一直努力向景观中灌输的超脱感和宽敞感。广阔的视野形成的宁静与和谐的形态和活动区巧妙地融合在一起。

A strong ecological profile forms the foundation of the whole transformational process. Polluted grounds have been cleaned; asphalt and concrete have been retrieved and reused; new soil for cultivation has been made from masses from the site. Large volumes of earth and rock within the Fornebu area have been used to transform the flat airport area into a landscape with different spatial qualities and heights to create views towards the fjord. Engineering firm Norconsult and the German firm Atelier Dreiseitl were consulted during the planning phase.

强烈的生态概貌成为整个改造过程的基础。受污染的土地已进行清理，沥青和混凝土都进行了回收和再利用，用于培植的新土壤已从现场的土块中获取。Fornebu地区的大块土和岩石把平整的机场用地打造成了具有不同空间特质的、高度不同的景观，面对着海峡。工程公司 Norconsult 和德国公司 Atelier Dreiseitl 都参与了项目的规划。

A rigid urban airport strip

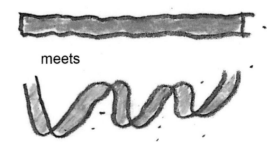

meets

a soft green naturalscape

Pilestredet Park

Pilestredet 公园

LOCATION: Oslo, Norway
项目地点：挪威 奥斯陆

AREA: 70,000 m²
面积：70 000 平方米

AWARD: The City of Oslo Architecture Prize (Oslo Byes Arkitekturpris) in 2005 and the National Building Design Prize (Statens Byggeskikkpris) in 2007
奖项：2005 年奥斯陆城市建筑奖；2007 年国家建筑设计奖

LANDSCAPE ARCHITECT: Bjørbekk & Lindheim
景观设计：Bjørbekk & Lindheim

Pilestredet Park

Pilestredet Park is an urban—ecology pilot project. When the old National Hospital in Oslo moved to a new site, more than 17.3 acres were converted to a residential and recreational area in the middle of the town. Pilestredet Park is a car free oasis in the city centre with vehicles largely directed outside an area that is designed to accommodate the needs of cyclists and pedestrians. Surface water drainage and storm water management characterise the facility and exploit the natural 16—meter fall of the site. There are rippling streams, water canals and pools in all outdoor areas. Every drop of water is taken care of and used several times to trickle, flow and drip, or lie perfectly still and reflect the sky and the treetops.

Pilestredet 公园是一个城市生态学试验项目。奥斯陆原国立医院迁址之后，城镇中心超过 17.3 英亩的地区都变成了居住和娱乐区。Pilestredet 公园是市中心一片没有汽车的绿洲，车辆大部分都安置在为骑脚踏车的人和行人设计的区域之外。地表水系和雨水管理体系以其设施为特色，并充分利用了场地 16 米的自然落差。室外有潺潺而过的小溪、水道和水池。每滴水都经过精心的处理并反复使用，流淌、滴落，或静止不动，反射着天空和树梢。

The project is based on the environmentally friendly principle of recycling building materials and elements from the old hospital. Stairs, foundation walls, window frames and granite gates have been preserved and reused in flooring, stairs and edging. The venerable portals have been used in the climbing wall or reset as part of the frame for the sandpit and pools. Concrete and other building rubble has been crushed and used for refill and as an aggregate in cast concrete used in the roads and public spaces. The terms "rag rugs" and "patchwork" from the world of textiles are thus directly transferred in the making of outdoor flooring in Pilestredet Park.

本项目遵守环境友好原则，使用了原医院的建筑材料和构件。原医院的楼梯、基墙、窗框和花岗岩大门都被保留了下来，重新利用到地面、楼梯和边缘的建造中。古老的入口门被用作攀岩墙，或作为沙坑或水池框架的一部分。混凝土和其他建筑石块都进行了破碎处理，然后重新用作回填或用于道路和公共空间使用的浇筑混凝土的骨料。在 Pilestredet 公园中，纺织业的术语"碎呢地毯"和"拼凑"被直接应用在室外地板的建造之中。

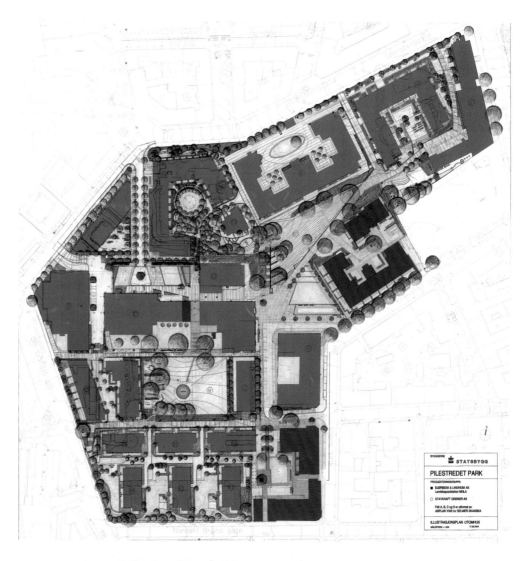

The old pool has been restored and a new element is added, which is a steel frame with a water curtain feeding water into the basin. This has become a popular playing area for children, as has the large arrow—head snake that children use in all sorts of ways.

旧水池进行了修复，并增加了一个新的元素，即一个带水幕的钢架，向池中注水。这里有一个大的箭头形弯道，孩子们可以以各种方式玩耍，因而这里成为深受孩子们喜爱的嬉戏场所。

When the entire hospital moved, the site was left with majestic, leafy trees and extra effort has been invested in retaining these, even in circumstances that they are very close to buildings. New trees have been planted to replace the old ones when they eventually die and a lot of bushes, creepers and ground-cover plants have been added to create a lush, green area in the centre of town. Ground-cover plants also reduce the need for constant attention to weeding and facilitate maintenance.

医院搬走以后，留下了茂盛、雄伟的大树。即使这些大树距离建筑很近，我们也设法将其保留下来。如果老树死掉了，我们会重新种植新的树木，同时我们还种植了许多灌木、爬行植物和地被植物，在市中心打造出郁郁葱葱的绿色景象。地被植物方便维护，无需经常除草。

The old National Hospital built in 1883 was surrounded by a massive wall to protect it from the public; this has been placed under heritage orders but punctuated in several places to give better access to the new area. Once inside, there are no fences or restrictions and the public can move freely from one outdoor space to another. These public spaces are used by many people as a pleasant green oasis on the way to and from work, whereas others use it as a place to escape from the turbulence of the city, a place to peacefully eat their lunchtime sandwiches or relax in a peaceful, green environment whilst still in the centre of the town.

原国立医院建于1883年，厚重的围墙将其与公众隔离。遵照遗产保护的指令，这些都被保留下来，不过有些地方也做了一些改动，以方便人们进入新建的区域。进入之后就没有围墙或限制了，人们可以从一个室外空间自由地进入另一个室外空间。很多人都将这些公共空间当作上下班途中舒适的绿洲，而另外一些人则将其做为远离城市喧嚣的场所，或是可以安静地吃午餐三明治的场所，抑或是在市中心的一处平和、绿色的放松场所。

Rolfsbukta Residential Area

Rolfsbukta 住宅区

LOCATION: Oslo, Norway
项目地点：挪威 奥斯陆

AREA: 54,000 m²
面积：54 000 平方米

COMPLETION DATE: 2008
完成时间：2008 年

ARCHITECT: Arcasa
建筑设计：阿卡斯

LANDSCAPE ARCHITECT: Bjørbekk & Lindheim
景观设计：Bjørbekk & Lindheim

Rolfsbukta
Residential Area

Rolfsbukta is a bay at the north—east end of Fornebu. It is one of the few points where residential areas on the site of the old airport have direct contact with the sea.

Rolfsbukta 是 Fornebu 东北部的一个海湾。古老海港上只有少量可以直接与大海接触的住宅区，而此处就是其中一个。

Very early in the project, a decision was reached to reinforce the relationship of the bay with the sea and consequently a canal was extended into the bay. To bring the water as close as possible to the public, we planned the inner 2/3 of the canal as a freshwater canal and the last 1/3 as a deep saltwater canal, connected by a waterfall between the two levels.

我们在项目的早期做出了一个决策，就是加强海湾与大海之间的关系，因此修建了一条通向海湾的水渠。为了尽可能地接近水，我们计划把水渠内部的 2/3 作为淡水水渠，而剩下的 1/3 作为深水咸水水渠，两部分由一个瀑布连接在一起。

The residential area, Pollen, surrounding the inner part of the fresh water canal, is already completed. A large pool, a pond with stepping stones, a wooden pier and a fountain frame two sides of the Pollen buildings. The water surface is enclosed by a poured concrete ramp that slopes down on one side, and on the other side stairs lead down to a 20-centimeter deep pool. The canal is surrounded by formal beds of ornamental grasses and willow trees framed by non-corrosive steel, the same material used in a custom-made barbecue. There are several seats surrounding the pool made of pored concrete with wooden covering recessed into the concrete. There is also a large platform of trees planted in gravel together with long tables and benches.

围绕浅水水渠内侧而建的住宅区 Pollen 已经完成。一个巨大的水池、一个有踏脚石的池塘、一个柁墩和一个喷泉构成建筑两侧的景观。水面的一侧由一个浇筑的倾斜混凝土坡道围合，另一侧有一些台阶，通往 20 厘米深的水池。水渠周围是观赏禾草和柳树形成的整齐匀称的河床，周围采用不锈钢框架，与定做的烧烤区使用的材料相同。水池周围有一些用浇筑的混凝土建造而成的座位，木质的铺面嵌入混凝土之中。砾石中还种植了一些树木，与长桌椅一起形成一个大平台。

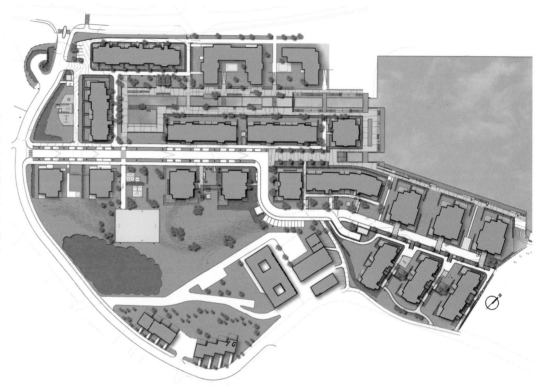

The seating, the "island" and the bridge
are lit from below, creating the impression
that they are floating as the dark falls.
More lighting is directed up towards the
trees from the planter boxes, throughout the
platforms, on the "island" and at the base
of the spray nozzles of the water fountain.

座位、"小岛"和桥从下面被照亮，夜幕
降临时它们仿佛漂浮在空中。种植箱、平
台、"小岛"和喷泉喷嘴的基座上都设置
了照明设备，直接照向树木。

Various species of willows and cherry trees
are planted close to the canal together with
the ornamental grasses.

水渠附近还种植了各种柳树和樱桃树，以
及一些观赏性植物。

Further out at Rolfsbukta, the second part of Phase 1, Tangen and Marina, have been completed. The pier motif with promenade is an essential part of this housing complex that is made up of six blocks on a north—west facing slope down towards the sea. You can moor boats and walk out along the bay to the outmost tip of the bay. This sunny west—facing waterfront is designed for recreation. Poured concrete embankments provide steps and seating.

在位于更远处的 Rolfsbukta 一期的第二部分，Tangen 和 Marina 也已完成。带有散步道的码头这一主题成为住宅综合体的基本部分，住宅综合体由六栋楼组成，位于一个朝向海面的西北向的斜坡上。你可以把船停泊在这里，沿着海湾散步，来到海湾的最远端。这个阳光充足的西向滨水区是为娱乐而设计的，浇筑的混凝土堤坝提供了台阶和座位。

Tjuvholmen, Oslo

Tjuvholmen，奥斯陆

LOCATION：Oslo，Norway
项目地点：挪威 奥斯陆

AREA：26,000 m²
面积：26 000 平方米

COMPLETION DATE：2010
完成时间：2010 年

LANDSCAPE ARCHITECTS：Bjørbekk & Lindheim, Jostein Bjørbekk, Rune Vik，Håvard Strøm，Sigrid Lofthus Drange，Aina Skjærvø，Christer Ohlsson
景观设计：Bjørbekk & Lindheim，Jostein Bjørbekk，Rune Vik，Håvard Strøm，Sigrid Lofthus Drange，Aina Skjærvø，Christer Ohlsson

Tjuvholmen, Oslo

Tjuvholmen，奥斯陆

LOCATION：Oslo，Norway
项目地点：挪威 奥斯陆

AREA：26,000 m²
面积：26 000 平方米

LANDSCAPE ARCHITECTS：Bjørbekk & Lindheim, Jostein Bjørbekk, Rune Vik，Håvard Strøm，Sigrid Lofthus Drange，Aina Skjærvø，Christer Ohlsson
景观设计：Bjørbekk & Lindheim，Jostein Bjørbekk，Rune Vik，Håvard Strøm，Sigrid Lofthus Drange，Aina Skjærvø，Christer Ohlsson

It is a transformation from harbour to maritime district.

这是一次从海港到海区的转变。

Tjuvholmen is one of the 13 projects in the Fjord City concept, Oslo's new waterfront, and is one of the most visible areas on the seaward approach to the central harbor area in Oslo.

Tjuvholmen，奥斯陆新滨水区，是采纳了峡湾城市理念的13个项目中的一个，也是通往奥斯陆中心港区、面向大海的通道上最显著的区域之一。

Tjuvholmen was originally one of the islands in the Oslo Fjord but was annexed to the mainland and incorporated into the dock and harbor area about 100 years ago. At Tjuvholmen, close contact with the sea will be re-established in a long coastline created by the new canals and public spaces and a lush park to be located along the shoreline. In addition to these opportunities for city life and recreation, there will be shops, restaurants, an activity centre with an art museum and a sculpture park to attract the public to Tjuvholmen, both in summer and winter.

Tjuvholmen 最初是奥斯陆峡湾众多岛屿中的一个，但是在一百年前与大陆合并在一起，成为港口和码头的一部分。Tjuvholmen 将沿着长长的海岸线重新与大海建立起密切的联系；这条长长的海岸线由新修的水渠和公共空间构成，海岸线旁有一个草木茂盛的公园。除了这些城市生活和娱乐设施之外，这里还将建立商店、餐厅和一个设有艺术博物馆和雕塑公园的活动中心；不论冬夏，Tjuvholmen 都将吸引大量的人群。

Tjuvholmen consists of three areasOdden, Holmen and Skjæret, and Bjørbekk & Lindheim are responsible for the concept and design of the landscaping of outdoor areas at Odden.

Tjuvholmen 包括三部分：Odden、Holmen 和 Skjæret。Bjørbekk & Lindheim 负责 Odden 室外景观的概念和设计。

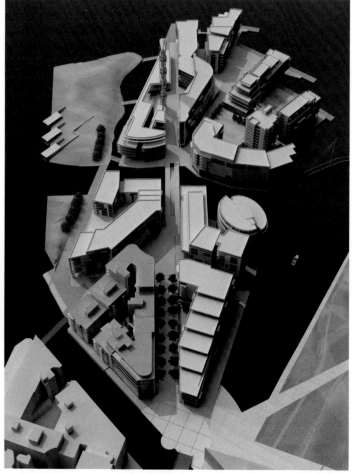

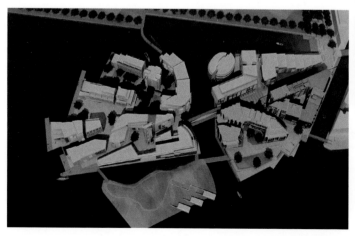

Above ground the district is car—free although there is parking for approx. 1000 cars in the underground parking garages. The street network of the district is laid out in a fan shape, creating changing patterns of sunshine throughout the day. The terrain on Odden has deliberately been created at the highest point at the central square. Three water features and a central tree create character with water features representing their different characteristics. Water runs from the fountains in channels through the streets towards the fjord.

地上是无车区，停车位都安排在地下，地下停车场中大约有 1 000 个车位。这里的街道网络呈扇形，在一天之中阳光在街道上不断变换。设计师刻意地将 Odden 置于中央广场的最高点。三处水景和中间的一棵大树突显出空间的特征，每处水景都有自己的特色。水从喷泉中通过渠道流向峡湾，贯穿整条街道。

Tjuvholmen is available for all users. Along the canal and quaysides, efforts have been made to create walkways that provide accessibility for all. There are plenty of seats and benches and a small amphitheater that provides seating and leads right down to the water surface.

Tjuvholmen 是属于大家的。人行道沿着水渠和码头而建，任何人都可以进入。这里有大量的座椅和长凳，还有一个小的阶梯式座位区，直接通向下面的水面。

Town Hall Park at Kjenn in Lørenskog

Kjenn 市政公园，Lørenskog

LOCATION：Akershus，Norway
项目地点：挪威 阿克什胡斯

AREA：39,000 m²
面积：39 000平方米

LANDSCAPE ARCHITECT：Bjørbekk & Lindheim
景观设计：Bjørbekk & Lindheim

Town Hall Park at Kjenn in Lørenskog

The park is in constant use with nursery schools and school children use it during winter days for skating and tobogganing on the ice rink and in summer for concerts, picnics, fly fishing courses, fishing, kiting, playing remote controlled boats and airplanes, feeding ducks and so on.

托儿所经常使用这个公园。冬天的时候，孩子们可以在溜冰场滑冰或者乘雪橇滑雪；夏天的时候，孩子们可以在这里举行音乐会、野餐、参加飞钓课程、钓鱼、放风筝、玩遥控船和遥控飞机，或喂鸭子。

The park is located close to the town centre in beautiful natural surroundings close to the Town Hall, to Lake Langevann, to the new Mainland High School and also to Kjenn Junior School. Lørenskog Municipality lies to the north of Oslo and is about a 15min commute. It is poised to carry out a comprehensive downtown expansion in which the large, new arts centre, "Lørenskog House", will be an important element. The new park is linked to the new Lørenskog centre via a new, soon to be opened pedestrian bridge over Route 159.

公园位于市中心附近，距离市政厅、Langevann 湖、新大陆高中和 Kjenn 初中都很近，周围自然风光秀美。Lørenskog 市政府在奥斯陆以北，来回大约15分钟。计划将进行一次全面的市区扩建，在此次扩建中，"Lørenskog 住宅" 是一个重要元素。新公园通过159大道上方即将开通的新建过街天桥与新的 Lørenskog 中心连接在一起。

With the new downtown development, the expansion of AHUS (Akershus University Hospital) and the jobs provided there and the establishment of a new large mail terminal for the Norwegian postal services within a few miles radius of the park, there is also a need for attractive green spaces, footpaths and bike paths that bind them all together. The town Hall Park at Town Hall will acquire a whole new status.

随着新市区的开发、AHUS（阿克什胡斯大学医院）的扩建及其提供的就业机会，以及距离公园几英里的挪威邮局大型邮政点的建设，这里还需要建设一些绿地、人行道和自行车道，使这些设施融为一个整体。市政公园将会具备一个全新的地位。

The Town Hall Park at the Town Hall is located in a well—established recreational area that has commonly been used for concerts, performances and as a recreational area for the city's inhabitants. Particularly in winter, the sloping grassy site has been an attractive place for tobogganing.

市政公园位于一个设施齐全的娱乐区，这里通常会举行一些音乐会和表演，市民也可以在这里娱乐，尤其在冬季，在倾斜的草地上乘雪橇滑雪是一个不错的选择。

Mark Klopferv

Mark Klopfer is a LEED accredited, registered landscape architect with seventeen years of practice experience. His experience is based on public building and park design, institutional and open space master planning, and on structure landscape. He has extensive public design process experience and has led numerous multi—disciplinary projects in both public and private sectors.

Mr. Klopfer is an Associate Professor of Architecture at the Wentworth Institute of Technology and has been a member of the architecture and landscape faculties at the Havard Graduate School of Design, Cornel University, the Rhode Island School of Design and Boston Architectural College. He was the 2000—2001 Prince Charitable Trusts Rome Prize winner at the American Academy in Rome, and currently serves on the editorial board of Architecture Boston.

Kaki Martin

Kaki Martin is a designer with thirteen years of practice experience. Her experience is based on public park design in most urban conditions, institutional and open space master planning, and river edge landscapes. She played an instrumental role in the design and construction of the North End Parks of Boston in Rose Kennedy Greenway. Ms. Martin has been an instructor in the core design curriculum at the Harvard Graduate School of Design and has taught at the Rhode Island School of Design. She is currently the Chair of the Cambridge Conservation Commission and is also a founding board member of Good Sports, a Boston—based non—profit organization serving disadvantaged youth.

Klopfer Martin Design Group

KLOPFER MARTIN DESIGN GROUP

It is committed to a design practice that sustains the balance between the built and natural landscapes. Our approach balances the aspirations of our clients with a respect for the needs of site, community, and sustainability. We have particular interest and expertise in urban sites, and the seam between architecture and landscape.

For more than ten years, Mark Klopfer and Kaki Martin have led complex teams that set new design vision for public landscapes. With current work in the North America and Asia, Mark and Kaki combine their design, public process and programming, and technical skills to create a team that is rich in aesthetic vision, experienced at project delivery, and stimulated by creative clients who face real-world, bottom-line constraints.

The Steel Yard

钢铁堆积场

LOCATION：Providence，USA
项目地点：美国 普罗维登斯

DESIGN COMPANY：Klopfer Martin Design Group
设计公司：Klopfer Martin Design Group

The Steel Yard

Most cities with industrial pasts inherit environmental problems in the future. The Steel Yard's cleanup is a showcase for regenerative design in a tough environment. Within industrial Providence, our project is a public intervention that upends commonly—held notions of blighted neighborhoods and shows the potential for real, actively engaged—not simply "adaptive"—re-use. The Steel Yard's landscape project embodies the non—profit (and necessarily inexpensive) mission through innovative brownfield remediation, stormwater filtration and reduction, purposeful design and placemaking.

大多数有着工业背景的城市在未来发展中都会存在环境问题。钢铁堆积场的回收利用会对这种艰难的环境起到至关重要的作用。在工业化的普罗维登斯，我们的主要阻力是老旧社区居民的普遍观念，我们要做的是不仅仅让他们去适应，更要让他们积极参与再利用过程。钢铁堆积场改革，是一个非营利性且开销不大的任务，但却能使污染得到整治，雨水得到过滤和还原，场所也会得到有目的的规划设计。

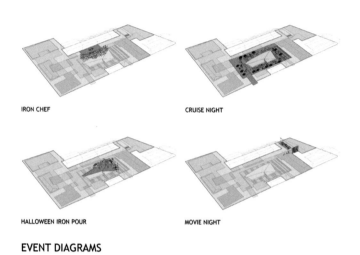

IRON CHEF

CRUISE NIGHT

HALLOWEEN IRON POUR

MOVIE NIGHT

EVENT DIAGRAMS

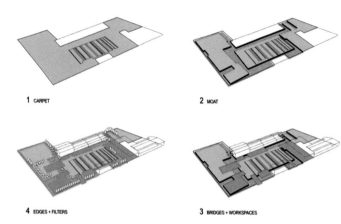

1 CARPET

2 MOAT

4 EDGES + FILTERS

3 BRIDGES + WORKSPACES

ORGANIZATIONAL STRATEGY

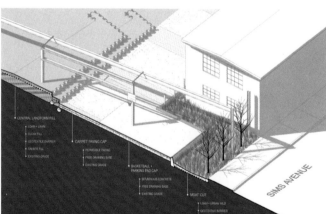

CAP SECTION

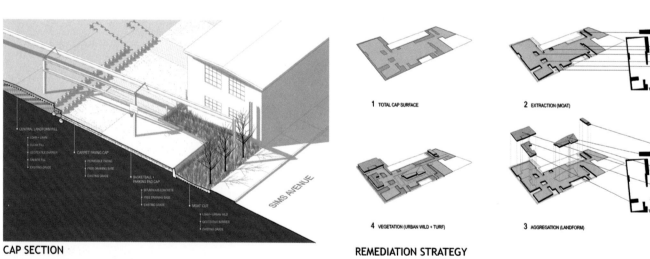

1 TOTAL CAP SURFACE

2 EXTRACTION (MOAT)

4 VEGETATION (URBAN WILD + TURF)

3 AGGREGATION (LANDFORM)

REMEDIATION STRATEGY

EXISTING CONDITIONS PLAN

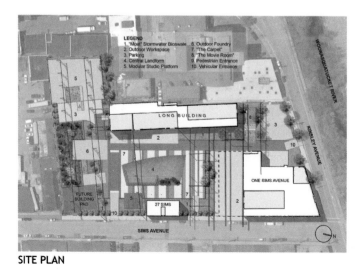

SITE PLAN

The Steel Yard acts as a catalyst in the creative industrial valley, district of Providence, Rhode Island. In fostering the industrial arts and incubating small business, the non-profit project seeks to cultivate an environment of experimentation and a community strengthened by creative networks.

钢铁堆积场在富有创造性的工业山谷——位于罗德岛州的普罗维斯登地区，起到了一种催化剂的作用，在鼓励工艺美术发展及扶持小型企业的过程中，这个非营利性的项目希望能发展出一个适应于试验的环境和一个由创新性网络强化的社区。

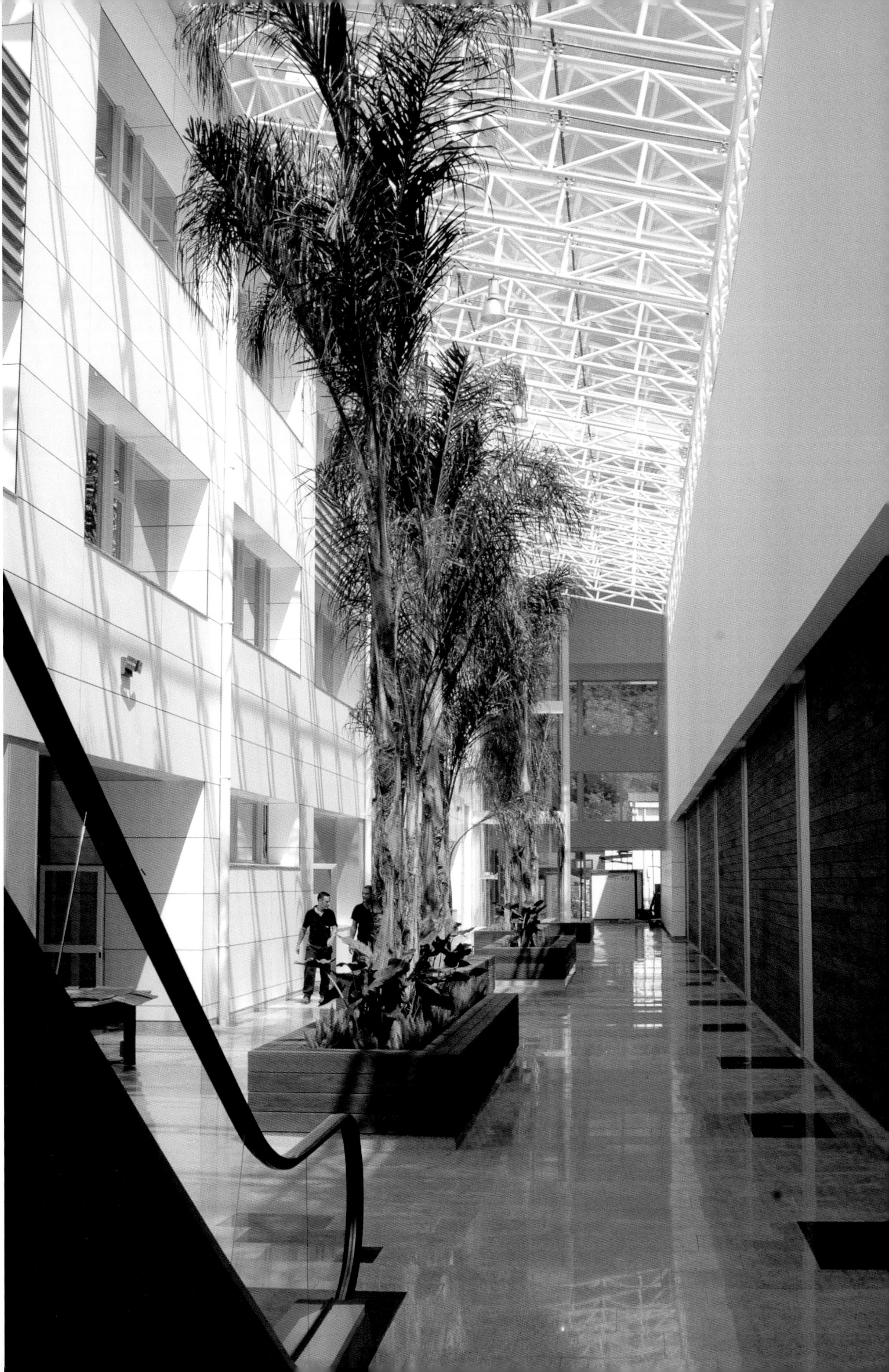

Iñigo Segurola Arregui

Education
1986—1989 Agricultural Engineering at the Public University of Navarra
1993—1995 Degree in Landscape Architecture from Heriot—Watt University ,
Edinburgh,
Scotland
1994 International Course on Planning and Landscape Design at the University of
Wagheningen, the Netherlands

Professional Activities
· Partner of LUR Paisajistak, S.L. since its foundation
· Address of the 3rd Workshop, specialty of Landscaping in the School of
Architecture, Pamplona
· Professor of 1° and 2° LOGSE Elements of Garden Design School of Design
Kunsthal of Irun, from the year 1998 to the present
· Collaborations, teaching Landscaping, with the Department of Urban Planning at the
Technical College of Architects of San Sebastian, Granada and Valencia
· Management, implementation and presentation of the gardening section of the
television program Decogarden in Tele 5 and TVE 2

Juan Iriarte Aguirrezabala

Education
1986—1991 Degree in Advertising and Public Relations from the Complutense University
of Madrid
1993—1995 Degree in Landscape Architecture from the School of Architecture at the
University of Edinburgh, Edinburgh, Scotland

Professional Activities
· Partner of LUR Paisajistak, S.L. since its foundation
· Professor of Ecology, Botany and Geomorphology projects for years 1 and 2 LOGSE
Elements of Garden Design School of Design Kunsthal of Irun, since the academic year
1998/99 to 2005/06
· Address and presentation of the gardening section of the TV Taste Home (the 6th)

iñigo segurola   juan iriarte

LUR Paisajistak, S.L.

What are we?
We are dedicated to landscaping, to the projects of parks and public
and private gardens since 1994 .
According to the characteristics and requirements of each project,
we collaborate with different studies. In the case of development
projects, we work closely with the architectural firm ABR +
Architects.

What do we provide?
Landscape provides a new way to understand the urban
environment of our cities: responding to social and functional
demands of each area under study, providing a vision consistent
with the aesthetic of the moment and adding to our
understanding and adaptation to projects of ecological processes.
Plants are living things and require suitable conditions and future
maintenance to thrive over time.
Our long-time experience and knowledge of the plant kingdom are
specific and their requirements make our projects generally greener
giving vegetation the due importance and therefore projecting from
sustainable parameters.

Who are we?
Limited partnership was formed by the landscape architects Iñigo
Segurola Arregui and Juan Iriarte Aguirrezabala. Both
Iñigo Segurola Arregui and Juan Iriarte Aguirrezabala are LUR
Paisajistak, S.L.´s solidary administrators.

Alai Txolo

Alai Txolo 公园

LOCATION：Irun, Spain
项目地点：西班牙 伊伦

AREA：30,000 m²
面积：30 000 平方米

PHOTOGRAPHER：LUR Paisajistak, S.L.
摄影师：LUR Paisajistak, S.L.

COMPLETION DATE：2009
完成时间：2009 年

DESIGNER：LUR Paisajistak S.L.
设计师：LUR Paisajistak S.L.

Alai Txolo

The old garden had been colonized by natural vegetation, creating a closed landscape, giving the sense of insecurity to residents. The park project studied the need to create an open space to provide the desired sense of urban safety.

公园原先是一大片天然植被，这种封闭的景观给居民带来深深的不安全感。该公园项目旨在打造出开阔的空间，给人们渴望已久的都市安全感。

It creates a large open space colonized by natural grass and planted grass and cherry trees that occupy the central park. In The central lawn is sorted hospitably towards areas that help to soften the slope through the walls that also serve as banks.

天然草地和栽种的草坪共同形成了此开阔空间，公园中心还栽植了樱桃树。中央草坪笑迎八方来客，它也让充当堤岸的高墙显得不那么陡了。

Along with the main access road to Irun, we establish a walk respecting the existing trees and plant many Himalayan palm trees that will provide identity to the park.

我们的设计让人们在通向伊伦市的主干道上，可以欣赏到树林美景，而且栽植的众多喜马拉雅棕榈树也会成为 Alai Txolo 公园的身份标志。

The more limited park space is defined by building a playground. The park incorporates into its planning the route of a bike lane.

在较为有限的空间内修建一个运动场。公园的整体设计中还包括一条自行车道。

Ametzagaina Park

Ametzagaina 公园

LOCATION：Donostia, Spain
项目地点：西班牙 圣塞瓦斯蒂安

AREA：38 hm²
面积：38 公顷

COMPLETION DATE：2010
完成时间：2010 年

DESIGNER：LUR Paisajistak, S.L.
设计师：LUR Paisajistak, S.L.

Ametzagaina Park

ACCESSES

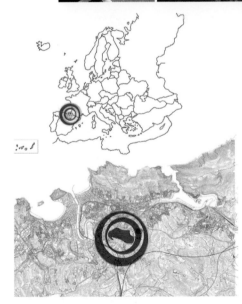

AMETZAGAINA PARKEA (DONOSTIA)

MASTER PLAN

Beneficiary:	DONOSTIA - SAN SEBASTIÁN City Council
investment:	8.500.000,00 €
Date:	Proyect 2005 - Realization 2008-2010
Location:	Donostia-San Sebastián (Gipuzkoa) Basque country
author:	Lur paisajistak s.l., abr+arquitectos s.l.
Area:	38 ha.
Typology:	Periurban forest park

LOCATION

The project's main action centres on the highest zone of the hill, the crest, where at present appears the ruins of a military fort and meadows. In this zone, we place the western viewing—point. The meadow and the fort are recovered as an enclosed garden with a resting area, the crater and the garden of the wind forming a chain of actions that allow to enjoy the park in its higher, sunny and horizontal zone.

本项目的主角是山的最高区域——山顶，现在山顶只有草地和军事堡垒的残骸。我们采用西方的视觉方式，草地和堡垒残骸整修后会营造出一个封闭的花园，休息区、火山口和风之谷又会形成景观序列，让人们在更高处、在和煦的阳光下和平坦处俯瞰公园。

VIEWING-POINT

The bulk of the project is completed with the layout of a network of pedestrian ways that cross and give access to the entire park. The crest's paths and the access to the peak possess an equal or minor slope of 6% to be accessible to Ametzagaina Park, even for the persons with physical handicaps. The path network is completed with the sports ones and the short—cuts.

整个公园纵横交叉的步行路径和通向四处的道路网是项目的主体部分。山脊小路和登顶的曲径的坡度接近，有的倾斜度甚至不足6%，身体有缺陷的人士也可顺利行走。道路网分为运动型和捷径两类，游人可自由选择。

THE CRATER

The entrances to the different paths of the park are proposed as squares of reception, composed of walls that on one side show "Ametzagaina Park" and on the other one contain informative panels of the path network and trails of the park, as well as the different resting areas and a naturalistic analysis of each zone.

通往公园的诸多道路入口处，拟建一个接待广场，广场上建两面墙，一面刻有"Ametzagaina 公园"字样，另一面则展示一些信息板，上面包括园内所有小路、曲径、各个休息区及每个区域的自然景色分析等信息。

In the center of the park, on the ridge between the fort and the crater, is constructed this resting area. This one is a space for rest, snack or games. In the tour of the Ametzagaina Park's crest, along with the Carlist Fort in the western part, the meadow is outlined. This is a built space quite open, capable of accommodating all kinds of activities or just walking. The fort is a romantic ruin that we turn into an enclosed garden.

休息区将建在公园中心、堡垒和火山口之间的山脊上，是一个集休息、饮食和游戏为一体的区域。去往 Ametzagaina 公园山顶的旅途上，道路左侧为卡洛斯堡垒，右侧则可看到草地的整个轮廓。这一处开阔的区域，既可作为各类活动的佳址，也可用于散步闲逛。废弃的堡垒摇身一变成了封闭的花园，真是妙不可言。

One of the points of major attractions of the park is the crater. It has been designed as a great concave space of grass with lawn at the bottom and meadow on the slopes where it is possible to enjoy shelter from the wind of any type and activities, from entertainment to sunbathing.

火山口也是该公园的主要景点之一。设计师将其改建为一个巨大的凹陷区域，底部是绿地和草坪，斜坡为无垠的草场，在这儿没有风的打扰，人们可以随心所欲地进行各种娱乐活动或者是享受日光浴。

RESTING AREA

CARLIST FORT

Jolastoki

Jolastoki 广场

LOCATION：Tolosa，Spain
项目地点：西班牙 托洛萨

AREA：6,785 m²
面积：6 785 平方米

COMPLETION DATE：2007
完成时间：2007 年

DESIGNER：LUR Paisajistak，S.L.
设计师：LUR Paisajistak，S.L.

Jolastoki

LOCATION：Tolosa，Spain
项目地点：西班牙 托洛萨

AREA：6,785 m²
面积：6 785 平方米

COMPLETION DATE：2007
完成时间：2007 年

DESIGNER：LUR Paisajistak，S.L.
设计师：LUR Paisajistak，S.L.

Plaza Jolastoki is located in Tolosa (Gipuzkoa, Basque Country) and the aim with this plaza is to create a thematic children playground that would gather kids and their keepers around a quality outdoor space. The centre of the plaza is designated to the playground with different kinds of games. Surrounding the central playground wooden decks are placed to enable a comfortable stay for the kids' keepers. All the periphery of the plaza is planted with different bamboos with punctual planting of cherry trees which add spectacular flowering during the end of the winter.

Jolastoki 广场位于西班牙托洛萨（巴斯克地区，吉普斯夸省），其设计宗旨是营建一个主题儿童游乐场，能够让孩子和他们的监护人相聚在高品质的户外空间，共享天伦。广场中央设定为游戏区，各种游戏项目应有尽有。在中央游戏区周围还设有很多木制台子，家长们可以舒舒服服地坐在台子上小憩。广场四周遍植翠竹，适时栽植的樱桃树在冬末会开出绚烂惊艳的花朵。

Orona

欧罗那

LOCATION: Hernani, Spain
项目地点：西班牙 埃尔纳尼

AREA: 1,400 m².
面积：1 400 平方米

COMPLETION DATE: 2008
完成时间：2008 年

DESIGNERS: LUR Paisajistak, S.L.
设计师：LUR Paisajistak, S.L.

Orona

欧罗那

Orona is a factory that builds lifts internationally and belongs to the world's biggest cooperative group (Grupo Mondragon). Located in Hernani, the factory has made a modernization process and the offices are growing, creating an indoor space for workers and visitors. The idea was to create an exuberant indoor space with palm trees and a pond. The arrangement of the plants and the pond was done by locating rectangles through the indoor space. The plants' periphery was designed wide enough to enable the workers to sit at it while having a break. The selected palm tree is the Arecastrum romanzofianum, which has developed a very good growth in these indoor conditions. The pond acts as an eye—catching feature to visitors and brings freshness and the sound of water to this indoor space.

欧罗那工厂隶属于世界最大的联合公司——蒙德拉贡集团，专门从事升降梯的制造，产品行销全世界。该工厂位于爱尔纳尼市，随着现代化进程的发展，工厂不断发展，已俨然成为工人和来访人员的室内空间。本设计旨在打造一个生气勃勃的市内空间，遍植棕榈，内有水池。植株周围留有足够空间，以便工人工作之余在此休憩。棕榈树则选择对室内环境具有很好适应习性的皇后葵。水池是吸引参观者目光的耀眼景点，潺潺的水声，为整个室内空间送来无限清新。

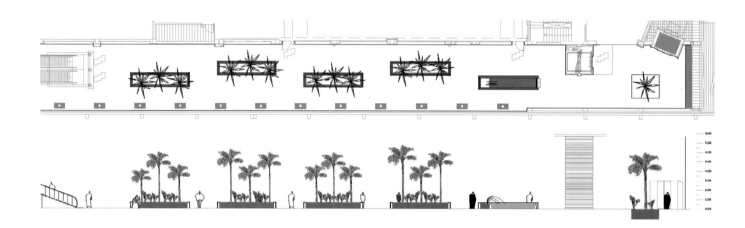

Plaza Europa

欧罗巴广场

LOCATION：Donostia，Spain
项目地点：西班牙 圣塞瓦斯蒂安

AREA：13,600 m²
面积：13 600 平方米

DESIGNER：LUR Paisajistak，S.L.
设计师：LUR Paisajistak，S.L.

Plaza Europa

Plaza Europa is located in one of the main accesses to the city of San Sebastian (Basque Country). The most singular feature in this plaza is the roundabout in which a sculpture of the artist Andrés Nagel is located. The idea was to locate the sculpture, a kind of horse with a lady on its back, under the canopy of eight pine trees. In this way the sculpture stays semi-hidden in the roundabout and the car drivers will discover it gradually. Besides the pine trees, the surface of the roundabout is modulated with small earthworks that bring a subtle relief to the space and this area is massively planted with ornamental grasses and perennials; Panicum virgatum, Miscanthus sinensis, Pennisetum spp., and Leymus arenarius are the main grasses and Rudbeckia sp., Dipsacus, Gaura lindheimeri, Onopordum cambricum and Leucanthemum sp. are the perennials that bring gradual flowering. The roundabout also has an artificial smog system that creates a magical atmosphere during the first and last hours of the day, while the light is soft and more people enter and leave the City of San Sebastian. This urban space won the first prize of Best Spanish Roundabouts in 2006.

欧罗巴广场坐落于一条通往圣塞巴斯蒂安市（巴斯克地区）的主干道上。该广场独领风骚之处就是环岛的设计，在那儿会竖起一座由建筑师 Andrés Nagel 设计的雕塑——在八棵松树的浓荫下，一位女士身骑骏马。雕塑在环岛中若隐若现，经过的司机们将会渐渐发现它们的存在。环岛的表面是一些小型的土方工程，给这个空间带来了精雕细琢之感，周围也遍植观赏性草和多年生植物，例如柳枝稷、中国芒、狼尾草、沙滨草等。草类植物以赖草属为主，黄雏菊属、川续断属、山桃草和滨菊属等多年生植物会在不同时令次第开放。环岛还设有一个人造烟雾系统，每日清晨和黄昏的两个小时，当光线渐渐柔和、人们进入或离开圣塞巴斯蒂安市时，人造烟雾会营造出一种朦胧神奇的氛围。本城市空间设计荣膺"2006年度西班牙最佳环岛设计"一等奖。

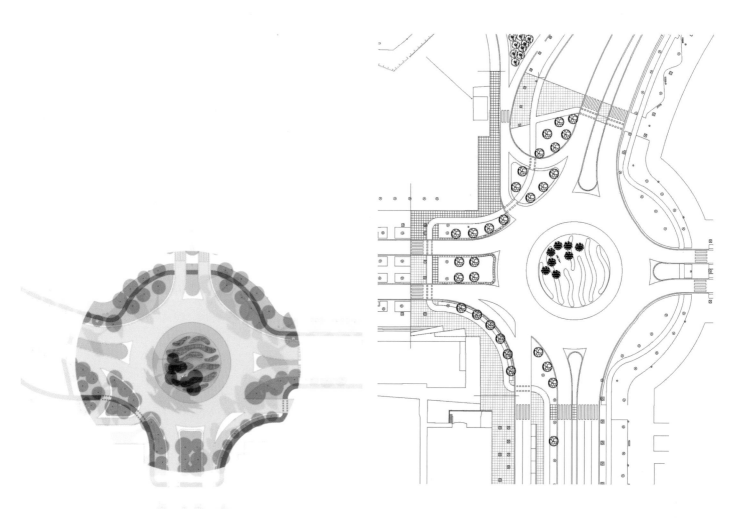

Riberas De Loiola

Riberas De Loiola

LOCATION: Donostia, spain
项目地点：西班牙 圣塞瓦斯蒂安

AREA: 2,712 m²
面积：2 712 平方米

COMPLETION DATE: 2008
完成时间：2008 年

DESIGNERS: LUR Paisajistak, S.L.
设计师：LUR Paisajistak, S.L.

Riberas De Loiola

Riberas De Loiola

Riberas De Loiola

LOCATION: Donostia, spain
项目地点：西班牙 圣塞瓦斯蒂安

AREA: 2,712 m²
面积：2 712 平方米

COMPLETION DATE: 2008
完成时间：2008 年

In the new neighborhood, Riberas De Loiola of Donostia buildings are grouped under the characteristic block model that characterizes the 19th century's urban expansion of Donostia. Within these blocks, public spaces, in the form of courtyards or patios, arise as a pedestrian— only use spaces.

Riberas De Loiola 坐落在圣塞瓦斯蒂安市一个崭新的街区环境，这个典型的街区模式融汇了 19 世纪圣塞瓦斯蒂安市城市扩张的各类特点。这些街区的公共地区是庭院和露台，只允许步行进入。

Paloma Plaza Miranda was designed, using large Cor—Ten steel containers as a construction unit. Each container has different measures and varies in height depending on its location, e.g. in areas which underground parking coincides with, planters are higher to provide more land to plant and allow further development. The boxes are distributed randomly, creating an interior space at the square used for a non—massive and calm purpose.

Paloma Plaza Miranda 计划使用耐腐高强度钢制成箱体，构成结构单元。每个箱的尺寸不一，高度不等，比如，在地下停车场周围要栽植很多高大的植物，箱子就要相对高些，为植株留出足够的空间，也备日后的发展之用。各箱子随意排列，隔出了一个奇特的内部空间，可供非大众性的、安静的消遣之用。

The vegetation selected is the Himalaya palm, Trachycarpus fortunei, the deciduous magnolia, Magnolia x soulangeana, Boxwoods are selected as bushes. The boxwoods in the future will be pruned to maintain the cubic forms of containers.

植被又分为乔木和灌木。乔木有喜马拉雅棕榈、其他棕榈树、落叶木兰、木兰属植物，灌木是黄杨树，日后将修剪黄杨树，让它们维持住箱体的正方形外观。

The black tile floor occupies center stage and obtains heat during the many rainy days that occur in the region, acquiring a special flair. The lighting during the day mimics the facades of buildings but at night takes on a special importance.

广场中间是黑瓷砖地面，本地区常年多雨，因而它能在雨天吸收热量，这就需要特殊的分辨力。在日光下，瓷砖会倒映出建筑物立面，到了夜晚则又具有了特殊的重要意义。

Villa Aitz Toki

Aitz Toki 别墅

LOCATION: Donostia, Spain
项目地点：西班牙 圣塞瓦斯蒂安

AREA: 13,500 m²
面积：13 500 平方米

COMPLETION DATE: 2010
完成时间：2010 年

DESIGNER: LUR Paisajistak, S.L.
设计师：LUR Paisajistak, S.L.

Villa Aitz Toki

The villa Aitz Toki is located at the northern slope of Mount Igeldo, Donostia, where you enjoy magnificent views over the Bay of Biscay. The garden is divided into two areas: the northern area overlooking the sea and the southern area. In the north, we just built a swimming pool with a large terrace to fully enjoy the views of the seascape.

Aitz Toki 别墅坐落在圣塞瓦斯蒂安 Igeldo 山北部斜坡上,在这儿你可将比斯开湾的壮阔风景尽收眼底。花园分为两个区域:可俯瞰海湾的北区和南区。在北区,我们仅建造了一座游泳池,配有一个大型平台,在这里你可以饱览美不胜收的海景。

In the south, which acts as access to the house, we created a simple but powerful garden, so the garden is an attractive contribution to supplement and partly to compete with the inherent appeal of the north zone. For this purpose, we chose to work with clean lines and to use Cor—Ten weathering steel as the unifying element of the whole.

南区是通往宅院的路口,我们设计了一个简洁而不失美丽的花园,使其成为了另一个吸引眼球的景致,并与北区相得益彰。为了实现此设计目的,我们选择了简洁的线条,并采用了符合整个设计风格的耐候钢。

To contain the gap with the neighboring estate, we created a retaining wall covered with Cor—Ten weathering steel which spouts out a pool, also built of Cor—Ten weathering steel, which empties into a linear pond built at ground level.

为了与邻近的宅邸留有距离,我们设计了挡土墙,表层为耐候钢。地面有一个线形水池,材质也是耐候钢。

From this retaining wall emerges a planter that shelters different plantations. This set provides visual simplicity and force that contributes to complement the breathtaking sea view of the north zone.

从挡土墙望去,可见一个种植园,里面栽种着各种各样的植物,奇景与北区令人叹为观止的海景相映成趣。

Owner
Dipl.—Ing. Till Rehwaldt, free—lance garden and landscape architect
Licence—No. Architect Association Saxony: 2553—93—4—c
1990 Diploma in Engineering (Dipl.—Ing.), Dresden Technical University (TU—Dresden)
1990—1996 Research assistant at TU—Dresden, the Landscape Architecture Department
since 1993 Free—lance landscape architect, focusing on object planning in public space
Expert consultant and judge
1999—2006 Lecturer at Zwickau University of West Saxony, Department of Architecture
2000—2002 Different university teaching positions at Dresden University of Applied Sciences, Department of Landscape Architecture
since 2001 Member of the Education Committee of the Saxon Architect Association
2003 German Landscape Prize 2003, Appreciation—Marien square Goerlitz
2006—2008 Visiting professor at Technical University Berlin, Institute of Landscape Architecture and Environmental Planning, Design Department
since 2006 Beijing Representative Office, China
2008 Appointment to the Leipzig Architectural Advisory Board
2009 German Landscape Prize 2009, first prize ULAP—square in Berlin
2009 Appointment into the convent of the Federal Foundation for Building Culture
since 2009 Appointment to the Executive Committee of the Federation of German Landscape Architects
since 2009 Teaching assignment "Open Space Development", Master of Urban Management program, Leipzig University

Employees
15 graduate engineers / landscape architects
2 technical employees

Working fields
Open space planning, object planning, conceptual open space planning, urban planning, revitalisation of housing areas, remodelling of mining landscapes

IT—Facilities
18 CAD—workstations, data transfers in all current formats, call for tenders, GAEB—compatible

Rehwaldt Landschaftsar-chitekten

Spaces Ahead

Modern Landscape Architecture is dealing with a site and its functional demands in a special way. The outdated concept of "style" has been replaced by focussing on the spatial and chronological context of a conceptual urban open space design.

For the newly designed space, a local identity is being created through the composition of urban open space and landscape as well as its tangible chronological context.

Regarding this, every space is a space for people. The public open space is organized in a multifunctional manner and offers possibilities for various uses. In this way the modern city becomes a human city. Neither representation nor decoration are dominating the future of urban open space but its quality as an everyday place of sojourn.

In 1993 our office was founded by the Dresden landscape architect Till Rehwaldt.

Main sectors of our scope of work are public open space planning, recreation and leisure facilities and urban planning. In our two offices (Dresden and Beijing), we are working on miscellaneous thematically and regionally diverse projects.

For us, taking part in design competitions is an ambitious recurring challenge. Most of our projects have been generated through this kind of planning process.

Bavarian State Garden Exhibition Burghausen 2004 — City Park

伯格豪森的巴伐利亚州园林展 2004 — 城市公园

LOCATION：Bavarian，Germany
项目地点：德国 巴伐利亚州

AREA：75，000 m²
面积：75 000 平方米

DESIGN COMPANY：Rehwaldt Landschaftsarchitekten
设计公司：雷瓦德景观建筑事务所

Bavarian State Garden Exhibition Burghausen 2004 — City Park

伯格豪森的巴伐利亚州园林展 2004 — 城市公园

LOCATION：Bavarian，Germany
项目地点：德国 巴伐利亚州

AREA：75，000 m²
面积：75 000 平方米

DESIGN COMPANY：Rehwaldt Landschaftsarchitekten
设计公司：雷瓦德景观建筑事务所

Within Burghausen's quarter Neustadt, an urban park has been developed as the core area for the "Landesgartenschau".

Right in the centre of the lively quarter, the relocation of a municipal building yard opened up the opportunity to completely redesign the urban open space.

Around an extensive meadow, differentiated garden areas have been developed. Their characteristics are corresponding to the neighbouring residential patterns and each garden features an independent formal special city.

For the first time, an interconnected urban open space system has been developed through the "Landesgartenschau". The City Park becomes an icon for the entire City of Burghausen. New walkway connections and view axes offer a hitherto unknown city experience. Consequently, the exhibition's activities are a crucial contribution to long—term urban development and creation of the city's image.

作为这次 2004 年"州园林展"的主要场地，在伯格豪森的新城建成了一座城市公园。三处不同边缘界定了一片宽广的草地并确定了公园的边界。这三个区域分别通过不同的用途和空间概念所区分。

在北面，"游戏山"是这个空间特有的元素。奇特的造型类似"微缩阿尔卑斯山"象征着对真实山峦的向往，它是服务于各个年龄人群的游戏场。"云雾森林"是一个特殊的空间，以独立的形式创立了对景观特性的参照。

通过此次州园林展第一次开发出一种城市开放空间的联系体系。城市公园成为了整个伯格豪森市的标志。新的人行道将其连接起来，视觉轴线创造了令人耳目一新的城市体验。展览活动也因此为城市的长期发展做出了贡献，并创造了城市的形象。

African Plains in the Dresden Zoo — Giraffes and Zebras

德累斯顿动物园里的非洲园 — 长颈鹿和斑马园

LOCATION: Bavarian, Germany
项目地点：德国 巴伐利亚州

AREA: 3,500 m²
面积：3 500 平方米

COMPLETION DATE: 2008
完成时间：2008 年

DESIGN COMPANY: Rehwaldt Landschaftsarchitekten
设计公司：雷瓦德景观建筑事务所

African Plains in the Dresden Zoo — Giraffes and Zebras

The imposing scenery of the park "GroBer Garten" is used as a visual extension of the zoo and so gives the giraffe bawn the impression of vastness. In elongation of the existing visitor ways, there are headlands extending in the outdoor enclosure. In this way the visitor can get the impression of being among the animals. The several zones of the bawn are assigned with topics like savannah, water hole or scrubland to show the giraffes in their natural habitat. The headlands give the opportunity to watch the animals at eye level.

For those who promenade in the park, the old zoo entrance is revived to the new "Zoo window". There are small keyholes in the portal where you can look through and get appetite of a visit in the Dresden Zoo.

长颈鹿园中令人难忘的风景作为动物园的视觉延伸，给人空间广阔的印象。除了传统的游览方式外，我们在户外设计了很多瞭望台，通过这种方式让游客感觉仿佛置身于动物之中。园区从几个不同主题来展现长颈鹿的自然栖息地，例如草原、水域、灌木丛。瞭望台提供了从人的视线高度观赏动物的机会。

对于那些在景观大花园中散步的人来说，动物园的老入口现在已经成为展现动物园风采的新窗口。公园入口大门上有很多小孔，通过这些观察孔人们可以了解到动物园里面的情况，从而产生参观德累斯顿动物园的欲望。

KaLa Playground and Green Space in Berlin — Friedrichshain

卡拉运动场和绿地 — 柏林弗里德里希斯海因

LOCATION: Berlin, Germany
项目地点：德国 柏林

AREA: 3,600 m²
面积：3 600 平方米

COMPLETION DATE: 2011
完成时间：2011 年

PHOTOGRAPHER: Rehwaldt Landschaftsarchitekten
摄影：雷瓦德景观建筑事务所

DESIGN COMPANY: Rehwaldt Landschaftsarchitekten
设计公司：雷瓦德景观建筑事务所

KaLa Playground and Green Space in Berlin — Friedrichshain

The urban quarters in the south of Frankfurter Allee are low—density areas with a high green rate ("green islands"). Small squares or green spaces accentuate the district and become sites with high amenity values. The design concept aimed to strengthen these characteristics and to enhance a distinctive identity.

这个城市小区位于法兰克福大道南侧，是一个低密度、高绿化率的小区（"绿岛"）。小广场或绿地为这个地区增加了色彩，使其环境舒适度大大提升。设计理念意在加强这些特性，增加场地的个性特征。

The opportunity arose to consider various target groups and different demands respectively for a lack in open spaces for children and elderly people. So, a great diversity of playing facilities for all age groups could be obtained in the quarters.

由于这里缺少供孩子和老人使用的空间，所以设计需要考虑不同的目标人群以及他们的不同需求。因此，在这个小区中随处可见供各个年龄段使用的各种休闲和娱乐设施。

The idea of two different sites—Meeresinsel (isle in the sea) and Erdeninsel (isle on the earth) came up to deal with the existing deficits. Through a well—balanced arrangement of the elements, disturbing interferences could be avoided.

为了弥补目前的不足，两个不同场所—Meeresinsel（海上岛屿）和 Erdeninsel（地上岛屿）的设计理念应运而生。通过对不同元素进行巧妙的安排，可以有效地避免冲突。

The Meeresinsel (Isle in the sea)—Play in
the current
The existing penguin playground has
been improved and refined in space and
issue. The sandy zone has been enlarged
southwards into a "big sea". The existing
playground equipment was maintained and
renewed. The motif of penguins, seals and
other sea dwellers was complemented by
new playing elements in the extension.
The "floes" are varied useable playing
objects which pick up the image of sheets
of ice which are drifting in the current.
Different topics characterise the floes and
create various playing situations: climbing
on the iceberg or walls, sliding or
balancing, "diving" or playing "marbles".
The surfaces are made of coloured concrete
which is partly coated. A flying cableway
is leading from one floe into a sandy area.

Meeresinsel（海上岛屿）—— 在水流中
玩耍
原企鹅运动场在空间和存在的其他问题方
面都得到了改善和提升。沙地向南扩建，
伸入"大海"之中。运动场原来的设施都
被保留下来，并进行了翻新。扩建中，企鹅、
海豹和其他海洋动物等主题与新游乐设施
相得益彰。"浮冰块"演变成各式各样的
游乐设施，使人联想到河流中漂浮的浮冰。
不同的主题赋予冰块不同的特色，打造出
不同的游乐场景：在冰山或墙壁上攀岩、
滑行或进行平衡游戏，"潜水"或"弹球"。
表面用部分涂漆的彩色混凝土建造。一条
架空索道从一块浮冰通向沙地。

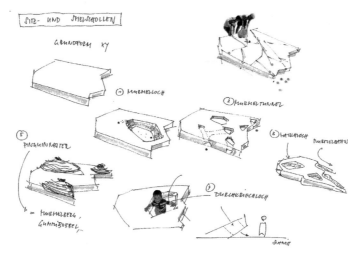

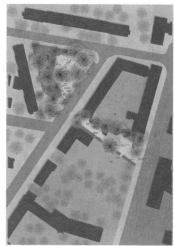

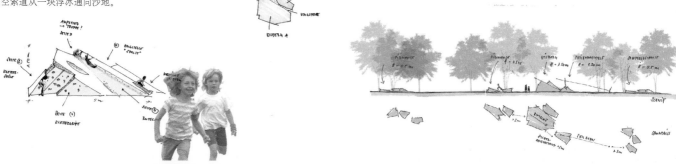

Jincheng Kinderpark

晋城市儿童公园

LOCATION：Jincheng，China
项目地点：中国 晋城

AREA：51,000 m²
面积：51 000 平方米

COMPLETION DATE：2010
完成时间：2010 年

PHOTOGRAPHER：Rehwaldt Landschaftsarchitekten
摄影：雷瓦德景观建筑事务所

AWARDS：Excellent Landscaping Project Gold Prize，China Society of Garden Landscape in 2010
奖项：优秀园林绿化工程金奖，中国园林风景协会 2010 年

DESIGN COMPANY：Rehwaldt Landschaftsarchitekten，Dresden，Germany
设计公司：雷瓦德景观建筑事务所德国德累斯顿分部

Jincheng Kinderpark

The Jincheng Botanical Garden (Kinder park) featured various potentials within its design before reconstruction. Mainly these were its valuable tree population, a Bonsai garden and a traditional bridge. The inaccessible waterfront without perceptibility, a non acceptable park entrance and the polluted East River were some of the topics that made it necessary to redesign the park. Views into the park from the outside and visual connections between park characteristics were important according to traditional Chinese gardens' needs.

晋城植物园（儿童公园）虽然有一些不足之处，但是经过新的设计，还是有许多可挖掘的潜力。首先，现有的大树十分具有保留价值，另外，盆景园和老桥也很有保留意义。尤其需要重新布置的就是湖岸区域，现在的湖岸缺乏亲水性和可体验性。没有得到重点体现的入口区和受到污染的东河都需要进行改造。因此，需要对各个不同的功能区进行适当的保留和重新设计。
按照中国园林艺术的传统，由园外向园内的视觉关系以及公园内部各景点之间的视觉关系是十分重要的。

The functional and formative trisection of the area in north—south direction was one of the main design ideas.

In the south there is an urban character that reflects modernity. It became a distinctive place as contemporary entrance and representation space.

The north is more landscaped and emblematizes tradition. This area is a recreation zone.

The interspace—the exhibition and event zone in the park center—is intended to convey information between urbanity and landscape. Therefore, design elements of both parts were set in content. Information boards on urban development and modern works of art were aligned.

Structures like the Bonsai garden and the Muslim cemetary which were worth retaining were integrated into the new design concept.

Besides the main entrance in the south, there are now more park entrances in the east and north.

The Jincheng surroundings such as islands, boardwalks, colourful and species—rich forests, lakes gave the inspiration for the new park design. In form, any natural landscape in China gave the model for the landscape park. The human—influenced agricultural landscape with its organic shapes in Jincheng region was used to design the exhibition area. The idea for the urban plaza derived from the aligned Jincheng urban structures. Three different park zones emerged by transferring the motives to the park site plan.

基本思路是根据设计和功能上的特点将园区由南至北分为三个区域。

南面的区域具有城市特点，并且反映了现代特点。这一区域，作为具有现代特点的入口空间以及着重表现的空间，为公园创造出独特的"地址"。

与之相对应，北面区域更多体现了自然风格，象征着传统。这里的主要功能是供人们休息。

中间的区域——展览及活动区——介于城市风格与自然风格、现代与传统之间。此区域的主题由展览的内容决定（介绍城市发展的信息墙和现代艺术品），形式上通过两个区域中的设计元素来实现。

需要保留的部分，如盆景园和回民墓，将融合在设计之中。

除了南面主入口，东面和北面也设有公园入口。

公园设计的出发点源于晋城的周边环境——岛屿、栈桥、颜色与植物种类丰富的森林、湖泊。所有的中国自然景观都可以成为这座景观公园在形式上的范本。晋城当地人为形成的起伏的农业景观激发了展览区的设计思路。城市广场的设计思路则是来自于晋城的城市设计。通过公园中不同区域之间的过渡形成了三个不同的公园区。

Redesign Southern Tan Mill Island (Südliche Lohmühleninsel)

南 "制革厂" 岛 (Südliche Lohmühleninsel) 重新设计

LOCATION: Berlin, Germany
项目地点: 德国 柏林

AREA: 65,000 m²
面积: 65 000 平方米

COMPLETION DATE: 2010
完成时间: 2010 年

PHOTOGRAPHER: Rehwaldt Landschaftsarchitekten, Dresden, Germany
摄影: 雷瓦德景观建筑事务所德国德累斯顿分部

DESIGNER: Rehwaldt Landschaftsarchitekten, Dresden, Germany
设计师: 雷瓦德景观建筑事务所德国德累斯顿分部

DESIGN COMPANY: Rehwaldt Landschaftsarchitekten
设计公司: 雷瓦德景观建筑事务所

Redesign Southern Tan Mill Island (Südliche Lohmühleninsel)

南 "制革厂" 岛 (Südliche Lohmühleninsel) 重新设计

LOCATION: Berlin, Germany
项目地点: 德国 柏林

PHOTOGRAPHER: Rehwaldt Landschaftsarchitekten, Dresden, Germany
摄影: 雷瓦德景观建筑事务所德国德累斯顿分部

Southern Tan Mill Island (Lohmühleninsel) is located in the quarter Kreuzberg, Berlin. The island evolved from the construction of the state canal (Landwehrkanal) during 1845 and 1850. It is 600 m long and 100 m wide. The tan mills which were formerly located on the site gave the island its name.

Most reconstructions of the public open spaces on Southern Tan Mill Island were finished in the years of 2007 and 2008. The new design is a good addition to the existing areas for a greater variety of activities for all age groups.

南"制革厂"岛（Lohmuehleninsel）位于柏林市克洛伊茨贝格区，由1845到1850年间建造的国家运河（Landwehrkanal）演变而来，长600米，宽100米。这个岛屿以原来位于这一地区的"制革厂"命名。南"制革厂"岛公共空间的大部分改建工程都于2007年和2008年完成。新设计为原来的区域增加了色彩，为各个年龄段的人群提供了各种活动设施。

The Southern Tan Mill Island is a popular recreation area and a frequently-used transit space for pedestrians and cyclists at the same time. For avoiding the subsequent functional conflicts, a guideline for the whole Southern Tan Mill Island was developed.

南"制革厂"岛不仅是一个受欢迎的娱乐区，同时还是一个行人和骑脚踏车的人频繁使用的交通区。为了避免日后发生功能方面的冲突，相关人员为整个南"制革厂"岛制定了设计指导原则。

Considering the overall situation, new intensive uses on the island were to be limited. Every additional activity area would have caused more usage frequencies and therefore further potential conflicts. So, it was important to well arrange additional recreational offerings sparingly and reasonably.

从全局考虑，新设计要对岛屿的集中使用进行限制。每个新增加的功能区都会被频繁使用，并可能导致冲突。因此，要分散、合理地布置新增设的娱乐设施。

Additional planning targets were the improvement of paths connections, the functional redesign of entrance areas, the better connection with Goerlitz Park and the establishment of visual relations. For a better orientation and the ordinary safety, shrubs and bushes were pruned. Additional plantings were placed only in selected areas.

其他规划目标包括改善小路之间的联系；对入口区域重新进行功能设计；与Goerlitz公园之间建立更好的联系；建立视觉联系。由于朝向和安全问题，对灌木丛进行了修剪。其他植物只能在选定的区域种植。

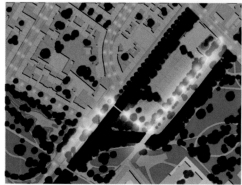
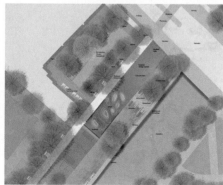

The wide main path as well as the playing strip were formed as generous active and transit spaces focusing on deceleration. The demands of different users are reflected on surface materials and features arranged on the playing strip. Cyclists, skaters and pedestrians share the spaces in slow motion or just pass by on the fast track in the northwest.

宽阔的主干道和条形娱乐区形成了一个宽敞的活动和交通空间，减速成为这里的主题。不同人群的需求反映在条形娱乐区的表面材料和特征上。骑脚踏车的人、滑冰的人和行人都可以在这里缓慢通行，或从西北侧的快速车道上通过。

Benches for resting are arranged along the main path. A large staircase partly equipped with wooden seats leads down to the channelside walk and can be used as a place for sojourn.

主干道的旁边设置了长椅，可以用来休息。一个巨大的台阶通向水渠边的人行道，上面设有木制的座椅，人们可以在这里驻足停留。

ULAP Square, Berlin

国家展会（ULAP）广场，柏林

LOCATION：Berlin，Germany
项目地点：德国 柏林

AREA：13,000 m²
面积：13 000 平方米

COMPLETION DATE：2008
完成时间：2008 年

PHOTOGRAPHER：Rehwaldt Landschaftsarchitekten，Dresden，Germany
摄影师：雷瓦德景观建筑事务所德国德累斯顿分部

DESIGN COMPANY：Rehwaldt Landschaftsarchitekten，Dresden，Germany
设计公司：雷瓦德景观建筑事务所德国德累斯顿分部

ULAP Square,Berlin

In 1879 an exhibition area was established in the west of the ancient Berlin between Lehrter Bahnhof, Alt—Moabit and Invalidenstraße. The site's characteristic was the division into two areas by the still existing commuter railway viaduct.

1879 年，在柏林的西部，柏林中央火车站，Alt—Moabit 和 Invalidenstraße 之间一个展览区被建立。展览区最显著的特征就是被一座通勤铁路高架桥一分为二。这座铁路高架桥至今仍被保留着。

Although planned as temporary arrangement for an industrial exhibition, the area was shortly developing to a permanent showground in Berlin named Universum—Landes—Ausstellungs—Park (Universe—National—Exhibition—Park), ULAP for short. Until short time before World War I, most of the big exhibitions of trade, hygiene, art and technics took place here.

这个展览区最初是作为一个临时工业展览区进行规划的，但是很快就发展成为柏林市一个固定的展览会场，并被命名为国家展览会场，简称 ULAP。直到第一次世界大战前不久，大部分大型的商品交易、卫生、艺术和技术展览会都在这里举行。

The first show was the trade exhibition in 1879, which was because of its innovative products the precursor for subsequent big industrial exhibitions. It also marked Berlin's beginning as a great exhibition city.

第一次展览是 1879 年举行的商品交易展览会，会上新颖的产品为以后大型工业展览奠定了基础，从那以后，柏林逐渐发展成为世界重要的展览场所。

For this reason, two exhibition buildings were erected on each side of the railway viaduct, of which the archs were also used as showground. The railway line was not operating by then. The planner of the site had to deal with a 4 m height difference to the adjacent street level.

为此，铁路高架桥的两侧各建起了一座展览大楼，高架桥的拱形结构也被用作展览场地。那时这条铁路线还没有开通。场地的规划者需要解决场地与附近街道之间的 4 米高差的问题。

At that time, the main entrance of the exhibition area was situated on the street Alt-Moabit. Via a wide two-way staircase featuring a water cascade, the visitor attained the main building.

那时，展览区的主入口位于 Alt-Moabit 大街上，游客可以通过一个宽阔的双向楼梯来到主建筑，小瀑布成为楼梯的亮点。

From there, people attained a forecourt with water basin and flower beds which also integrated the main staircase. The site in the west of the railway viaduct was landscaped. Besides the single pavilions and facility buildings, a large artificial lake formed the centre of the fairground.

从这里，人们可以进入一个前院，前院里面有水池和花坛，与主楼梯融为一体。铁路高架桥的西侧进行了景观美化。凉亭与设施楼之间有一个大型的人工湖，成为会场的中心。

After the devastating fire of the exhibition hall, shortly before the opening of the General German Exhibition for hygiene and rescue in 1882, an intensive discussion about the future of ULAP emerged. Under the pressure of an expected universal exhibition in Berlin, possible alternative fairgrounds were discussed. Even the usage of Lehrter Bahnhof was considered, which was planned to shut down at that time.

在一场毁灭性的火灾之后，同时在 1882 年建立德国卫生和救援展会前不久，人们对 ULAP 的将来进行了大量的讨论。由于柏林将要举行一次世界展览，因此有人提议在这里建一个交易会场，甚至考虑到了使用当时即将关闭的柏林中央火车站。

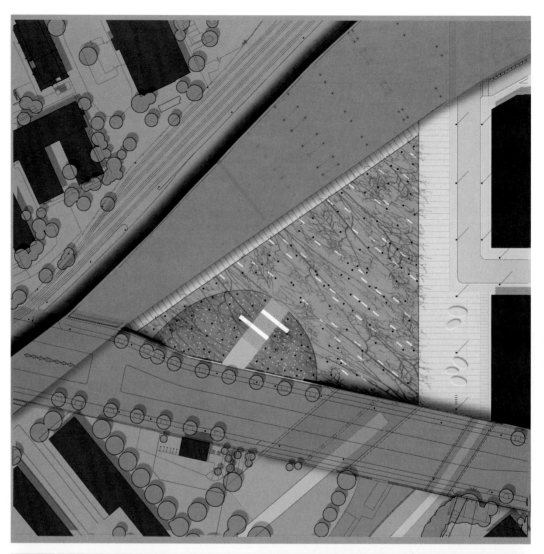

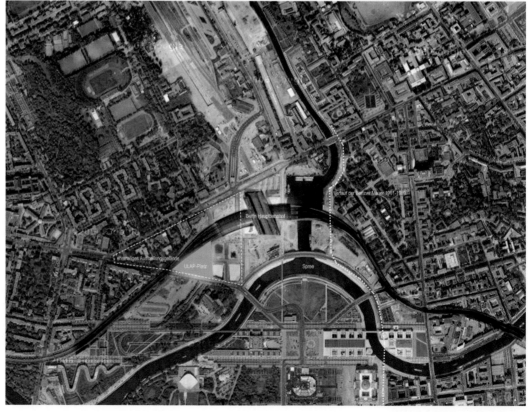

Nature in Waldkirchen 2007 — A Decentralised Exhibition Concept

2007 Waldkirchen 花园展的本质 — 分散展览理念

LOCATION: Bavarian, Germany
项目地点：德国 巴伐利亚州

AREA: 75,000 m²
面积：75 000 平方米

COMPLETION DATE: 2007
完成时间：2007 年

LANDSCAPE DESIGNER: Rehwaldt Landschaftsarchitekten, Dresden, Germany
景观设计：雷瓦德景观建筑事务所德国德累斯顿分部

DESIGN COMPANY: Rehwaldt Landschaftsarchitekten, Dresden, Germany
设计公司：雷瓦德景观建筑事务所德国德累斯顿分部

Nature in Waldkirchen 2007 — A Decentralised Exhibition Concept

This small garden festival is based on a decentralised exhibition concept. The main area of the garden exhibition is designed as a loop path from the core of the garden exhibition, the new Waldkirchen urban park, via old town to the characteristic areas in and near Waldkirchen.

本次小型花园节以分散展览理念为基础。花园的主展区被设计成环形路线，从花园展的核心——新 Waldkirchen 城市公园，途经老城区，到达 Waldkirchen 市内和附近的特色区。

As the main feature of the garden exhibition, the new urban park was developed on the south-eastern edge of the old town, along brook Waeschlbach. Through reorganisation of existing streets, parking sites and a bus stop, a spacious entrance square to the urban park and the garden exhibition could be established. Here, all main events and activities can take place. "Landschaftsbalkone" (scenic balconies) are offering special views into the surrounding landscape and are explaining them.

新城市公园位于老城区的东南侧，Waeschlbach 小溪沿岸，成为花园展的主要特色。对原有街道、停车区和公共汽车站进行重新规划之后，有必要在城市公园和花园站附近建一个宽敞的入口，所有的重大事件和活动都可以在这里举行。观景台提供了观赏周围景色的特殊视野，同时也对周围的景色进行了很好的诠释。

The Waeschlbach valley features a variety of scenic and historic sites like the natural landmark "Gsteinet", floodplain forests and hillside meadows along Waeschlbach, as well as some protected habitats in an urban context. Along the elongated meadows, the manifold topographies and the diverse landforms were used to develop differentiated gardens and to provide great views into the landscape. Within the garden exhibition, a new open space system was established, which links the urban park and the adjacent residences through new paths.

Within the structural facilities and design focuses, the Waldkirchen's characteristic location as part of the Bavarian Forest was respected. Therefore, the material wood plays an important role as functional and artistic elements. The elements of the urban park such as entrance square, urban promenade, water stair case, cherry gardens, pond and Waeschlbach valley as well as the spacious meadows with the "Gsteinet" are understood and designed as self-contained areas. Manifold views beyond the park borders enlarge the spatial impression of the park.

An additional focus of the garden exhibition is the garden "Bellevue"along the loop path. It represents the city's character besides the other municipal open spaces such as market-place, cemetery, urban park and sports fields.

At the "Bellevue", the unique view at Waldkirchen was used and staged through a self-contained design. Waterbound paths frame the shrub and bush plantings and show the inside and outside areas. On a spatial view point, a water fountain was placed which has playground and recreational zone in one. On historic site of Waldkirchen's water reserve, the site was newly interpreted by using instruments of contemporary landscape architecture.

Waeschlbach 溪谷以各种景点和古迹著称，如大自然的地标"Gsteinet"，Waeschlbach 沿岸的泛洪区森林和山腰草地，以及城市中一些受保护的生态环境。设计利用长长的草地旁边多样化的地形打造出不同的花园，提供了绝佳的观景视野。花园展内建立了一个新的空地系统，通过新建的小路将城市公园与附近的住宅联系在一起。

结构设施和设计焦点都尊重了 Waldkirchen 作为巴戈利亚森林一部分的独特位置。因此，作为功能和艺术元素，木材在这里发挥着重要作用。城市公园的元素，如入口广场、城市散步道、水景楼梯、樱桃园、水塘和 Waeschlbach 溪谷，以及带有 Gsteinet 的宽敞草原都被融入设计之中，组成一个设施齐全的区域。公园另一侧，远处多样化的景观增加了公园的宽敞感。

花园展的另一个亮点就是环形路旁边的"Bellevue"花园，它代表着除了市场、公墓、城市公园和体育设施等城市空间以外的另一个城市特征。

通过综合的设计，Waldkirchen 独具特色的景观在"Bellevue"重新上演。亲水小路与灌木相互呼应，显露出内部和外部区域。鉴于空间的因素，喷泉所在的位置既是一个运动场，又是一个娱乐区。作为一个以 Waldkirchen 的水资源为特色的古迹，场地通过当代景观建筑得到了重新诠释。

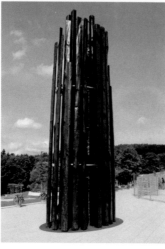

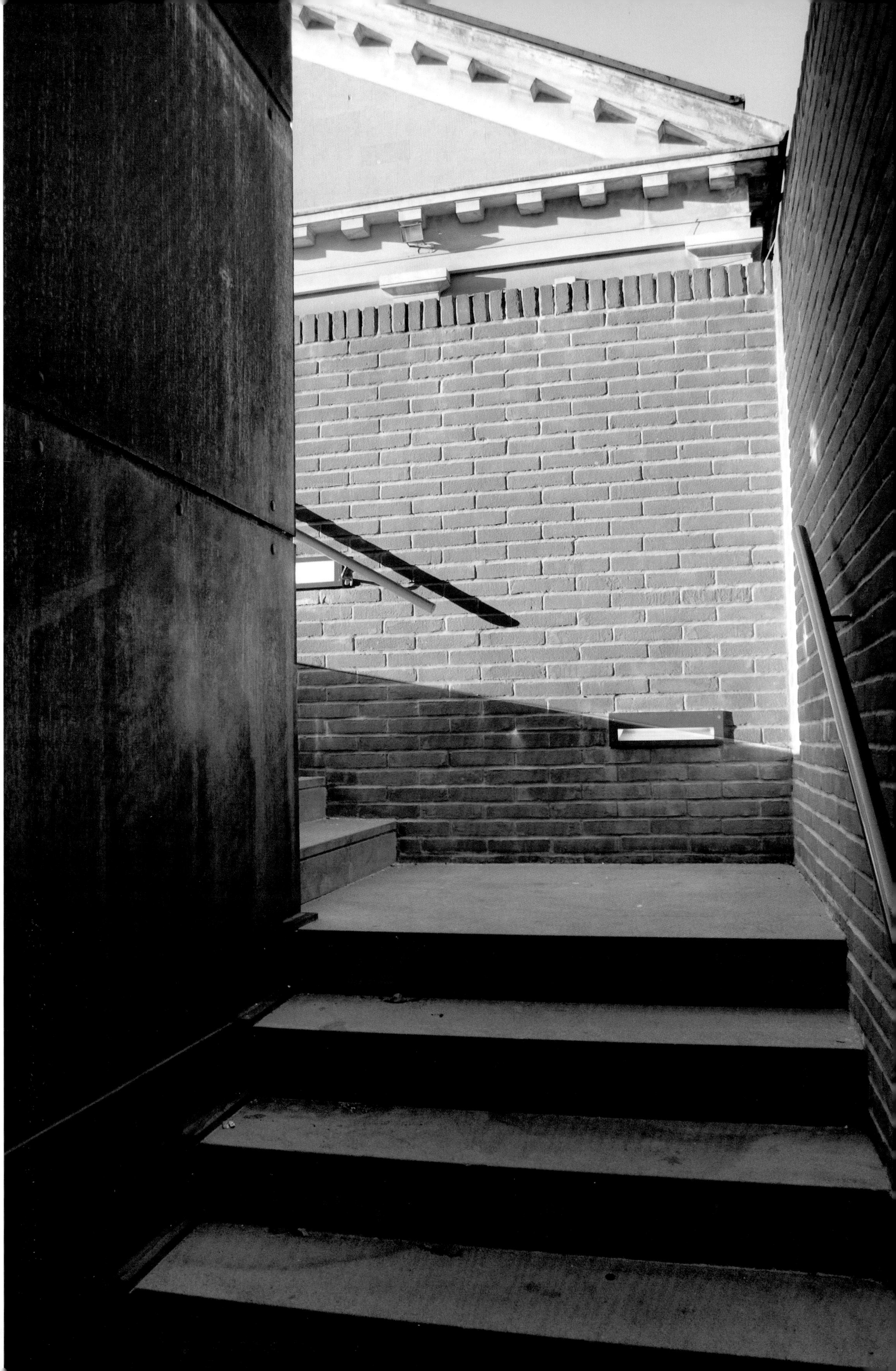

Gianfranco Franchi

Gianfranco Franchi was born in Lucca in 1954. He studied architecture at Florence
University and landscape architecture at the University of Genova.
In 2001, he established associated office Franchi Lunardini partners and in 2007
FRANCHI ASSOCIATI office in wich he is the owner and art director.
Since 2006, he is vice president of the Italian Association of Landscape Architecture.
Since 2001, he teaches as a professor of landscape architecture at Genova University.
He has lectured at a variety of universities and cultural institutions.

FRANCHI ASSOCIATI

FRANCHI ASSOCIATI

It is a landscape architecture studio that specializes in the design and
construction of urban open spaces and public parks and gardens.
The analysis of the sites, the search for conceptual approaches
means that we can define the right solutions for each context.
We have gained manifold experiences through a broad spectrum of
projects.
Each project such as new parks, urban spaces, gardens,
environmental restoration, strives to respond to site conditions and
programmatic necessities with a compelling concept, high quality of
design and efficient implementation.
The activity has spaced at every scale, from the general plan to the
private garden.
The office has completed many projects, mainly including:
waterfront of Savona, S. Anna park, Petroni square, Osservanti
park in Lucca, school spaces in Prato, east park masterplan in Lucca,
roof gardens, restoration garden in Lodi and other measures of
environmental sustainability.
The office has its disposal professional staff and partners who have
the necessary competences to carry out high quality projects
and to satisfy the needs of customers, who range from privates
to entrepreneurs, touristic operators, public administrations and
professional offices.
FRANCHI ASSOCIATI often collaborates with other professionals
specializing in engineering and technique parts to integrate the best
solutions during the project.
All construction drawings are done in-house to the highest
professional and environmental standards.

Giardino degli Osservanti

花园景致

LOCATION: Lucca, Italy
项目地点：意大利 卢卡

GARDEN AREA: 18,000 m²
花园面积：18 000 平方米

ROOF GARDEN AREA: 14,500 m²
屋顶花园面积：14 500 平方米

COMPLETION DATE: 2006
完成时间：2006 年

PHOTOGRAPHER: Gianfranco Franchi
摄影师：Gianfranco Franchi

DESIGNER/ARCHITECT: Gianfranco Franchi
设计师／建筑师：Gianfranco Franchi

LANDSCAPE ARCHITECT: Franchi Associati
景观设计师：Franchi Associati

Giardino degli Osservanti

The "Giardino degli Osservanti" is a public park realised within the 18th century walls of the city of Lucca. The aim of the project was to build an underground parking space for 530 vehicles allowing the creation of a green area for the citizens of the city. The project made it possible to retain a part of the city that would otherwise have deteriorated. This was achieved through the recovery of the 14th century monastery, the building of new dwellings and the use for academic activities.

这个"花园景致"是一个公园，建于卢卡市18世纪的城市围墙里。工程的目标是为530辆车建一个地下停车场，并且给市民创建一个绿色空间。该工程保留住了城市的一块区域，使它没有被荒废掉。这一目标的实现是通过恢复14世纪的寺院，修建新的住宅和为文化活动提供场所来实现的。

The design of the park was inspired by the traditional historic gardens of the city. For these purposes, historical research led towards the coinage of an alphabet of shapes, languages and materials that guided the project. The planimetric installation brings a contemporary reinterpretation to the typical structure of the gardens of the city, which used to stand within the monastery, during the Renaissance.
This new public space is linked to a policy of sustainable development and is integrated in the growth of the urban landscape.

公园的设计受到该市那些传统而具有历史意义的花园的启发。为了实现这些目标，历史研究涉及了曾经创造的图形和指导工程的各种语言和材料。平面设计给城市花园里那些典型的建筑结构一个当代的全新诠释。这些结构在文艺复兴时期曾坐落于那个寺院当中。
这个新公共空间坚持可持续发展政策，融入城市景观发展中。

The project retained a surface for the use of the public, separating it from the parking space and dedicating it to activities for social life and tourism.

该项目保留了一块供公众使用的场地，让它与停车空间隔开，为社交和旅游活动服务。

Moreover, a substantial importance has been given to the incrementation of green surfaces, improving the micro-climate and biodiversity.

而且，绿色场地的增加受到极大的重视，促进了小气候和生物多样性的发展。

The project, which alternates concrete spaces with green spaces, and presents lifted—up "parterre" and paths, is a clear reference to the characteristic internal division of the historical buildings of the city of Lucca.

这一项目把绿色空间和混凝土空间交替间隔开，呈现出凸起的花坛和小路，鲜明地体现了卢卡市历史建筑的内在特色。

The design originates from the mediation between these historical and artistic references and the lines dictated by the irregular orientation of the surrounding buildings.

既要做历史和艺术参考，又要考虑周围建筑由于不规则方位形成的线条，设计就是在两者的平衡中产生的。

The intervention involves an area of around 18,000 m², 14,500 m² of which are roof garden. The central area of the garden hosts the major elements of the project such as the lifted—up "parterre", the water games fountain and the sequence of hedges alternated with river pebble parterre.
The long arbours represent the natural mediation between the building and the garden and allow the enlargement of shaded spaces.

工程覆盖面积大约是 18 000 平方米，其中 14 500 平方米是屋顶花园。花园的中心区展现了项目的主要元素，比如凸起的花坛、水上游戏的喷泉、与河卵石花坛交替的一系列篱笆。那些长长的藤架是建筑和花园之间自然协调的体现，扩大了阴凉空间。

The trees introduced in the project are a type of the area: Morus Alba for the production of silk.
Access to the underground parking area is facilitated by stairs and lifts covered with Cor—Ten steel, which recalls the presence of the ex—military headquarters.

项目中采用的树是当地的特色植物：生产丝绸的桑树。
地下停车场设置有方便进入的楼梯和电梯，由耐腐蚀高强度钢遮盖着，让人回想起这是前军事总部的所在地。

During the realisation of the garden, an archaeological area has been discovered and has been encompassed in the design of the garden.

在花园的施工中，一个考古区域被发现了，并被纳入花园的设计中。

Giardino del Passeggio

漫步花园

LOCATION：Lodi，Italy
项目地点：意大利 洛迪

AREA：35,000 m²
面积：35 000平方米

COMPLETION DATE：2009
完成时间：2009 年

PHOTOGRAPHER：Gianfranco Franchi
摄影师：Gianfranco Franchi

DESIGNER：Gianfranco Franchi
设计师：Gianfranco Franchi

LANDSCAPE ARCHITECT：Franchi Associati
景观设计师：Franchi Associati

Giardino del Passeggio

The pathway of Lodi (1934—1949) is a long tree—lined area, structured as a formal garden at the entrance of the city, which appears to those who access the centre through the main roads. Its unique shape and superb location make it the gateway to the city and give the garden a spectacular and important scenic role together with the function of green area available to the public.

洛迪步行路（1934-1949）实际是一个正式的花园，面积狭长，周围绿树葱葱，位于城市的入口处，对于那些要穿过主干路去市中心的人来说，它会首先映入眼帘。它造型独特，地理位置极佳，不仅成为了进入城市的通道，还是一个风景秀丽的重要景点，为公众提供可利用的绿色空间。

The difficulties in the design included the intervention in a garden built in the 20th century in order to meet the current needs of very different cultural patterns. The garden was characterised by a strong and repetitive geometry of the design with rows of trees and an important "parterre" enclosed by hedges which precluded access to it.

设计上的困难包括按照现在的需要对一个建于20世纪的公园进行改变，使其满足不同文化模式。花园的特点是具有统一规则的几何图形，有成排的树和大花坛，花坛由树篱围着，阻止人进入。

We thought we could draw something very interesting, and that was what we did. We reinterpreted the forms and we built a new garden starting from the existing one. This process was incredibly complex and characterised by many discussions within the group of designers. It was hard to realise what would be the right balance between the old and the new design.

我们认为可以加点有趣的元素，然后我们做到了。我们对外观形式做了重新诠释，在现有的花园旁又修建了一个新的花园。这一过程极其复杂，设计小组内部进行了大量的讨论。把新的和以前的设计有机地结合起来是很困难的。

During the process of renewal, we identified new areas and activities that would be adherent to the needs of modern life. We introduced shaded areas as well as meeting and resting places.

在重新修建的过程中，我们依据现代生活的需要给各块区域赋予新的身份和功能。我们引入大量的阴凉处及聚会和休息的地方。

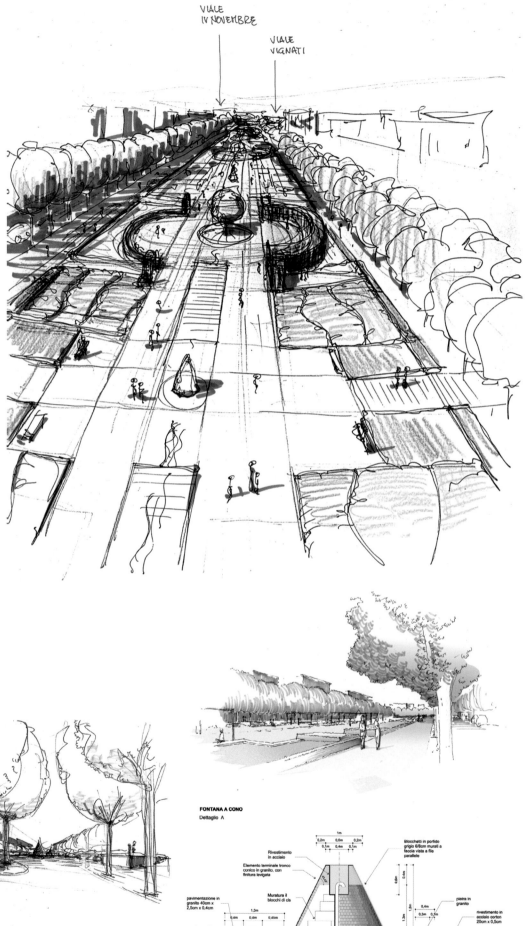

VIALE IV NOVEMBRE

VIALE VIGNATI

We decided to bring back water to the garden. Before the garden was built, water used to surround the entire city and we felt it was important to give it a new significant role.
However, the water used to flow at a very low level and we were faced with the question of how to reproduce something similar.

我们决定把水带回花园。在花园修建之前，水是环绕整个城市的，我们感觉让水扮演一个重要的新角色是很关键的。
然而，过去水流位置很低，我们面临的问题是怎样复制相似的景致。

Finally, water was introduced as the characterising element of the whole garden and it now runs through the various rooms that we designed. Its course starts from a fountain and becomes sinuous stone pavement. It then shifts back to the fountain creating water games. The water appears and disappears throughout the entire garden, generating curiosity and historical remembrance.
The choice of this approach was a happy one, because today the park is highly frequented by the people who are able to recognise themselves in the features of the garden.

最后，水作为整个花园的特色元素被引入，现在水流过我们设计的各种各样的空间。水流路线开始于一个喷泉，接着是蜿蜒的石面。然后它绕回到能做水上游戏的喷泉。水在整个花园里忽隐忽现，引起人们的好奇和对历史的记忆。
这个方案的选择令人感到高兴，因为今天人们络绎不绝地来到公园，他们在花园别致的景致中能够找回自我。

FONTANA A CONO
Dettaglio A

Sezione AA
scala 1:50

Unfortunately, this big turnout of people is creating some problems in the management and maintenance of the garden which are not easy to solve.

不幸的是，大量的游客给管理和维护方面带来了一些问题，很不容易解决。

An urban park, a garden, a landscape, should be developed according to the reading of existing structures, signs and history that each place has. I do not believe in an "international style" homologating the design of all the parks. It would mean loosing the characteristics of different places and the possibility of the "genius loci" to stimulate new forms and new uses. Only in this way can we try to experiment with and create new parks for the future.

一个都市公园，一个花园，一个景观应该依据每个地方存在的结构、符号和历史来开发建设。我不相信有一种国际风格能符合所有公园的设计。那将意味着为了激发新形式和新用途而丢失了不同地方的特色和树立"场所精神"的可能性。只有通过这种方式我们才能为了将来试着去实验并创造新的公园。

Orzali Park

奥尔扎利公园

LOCATION：Monsummano, Italy
项目地点：意大利 蒙苏马诺

AREA：38,000 m²
面积：38 000 平方米

COMPLETION DATE：2011
完成时间：2011 年

DESIGNER/ARCHITECT：Gianfranco Franchi
设计师／建筑师：Gianfranco Franchi

LANDSCAPE ARCHITECT：Franchi Associati
景观设计师：Franchi Associati

Orzali Park

The "Orzali Park" is a park situated in the suburb of Monsummano City, in Tuscany. It is close to the areas of urban and industrial expansion and it constitutes a link between the city and the countryside. The park is located along a small torrent: the Candalla torrent.

奥尔扎利公园坐落于蒙苏马诺市郊区，在托斯卡纳区。它靠近都市和工业扩张的区域，成为城市和乡村的一条纽带。公园位于 Candalla 河边。

We thought for a long time what would be the right design and activities to be included in the project of the park and which type of park would suit the most the needs of the environment. We opted for a non-urban park; a "wedge of neutrality" which enters in the built part of the city. The realisation of wide areas of trees will promote the improvement of the ecological aspects and will allow the shadowing of the area which is currently devoid of tree vegetation.

我们思考了很长时间，在这个公园项目中什么样的设计是适合的，应该加入什么样的活动，以及哪种类型的公园会最符合环境的需要。我们决定建一个非都市类型的公园；它像楔子一样插入城市建筑中，保持着中立性。里面要种植宽阔的树林，这样会促进生态方方面面的改善，还会给缺树木的地方遮光。

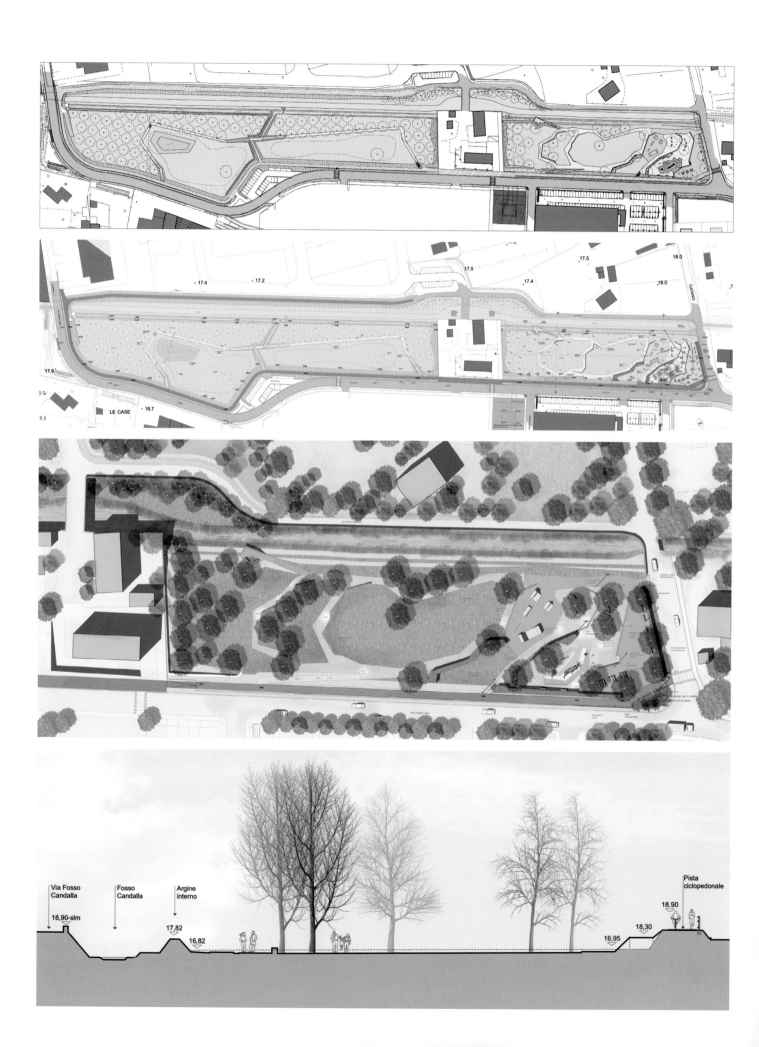

We divided the park into two areas: the first is an area dedicated to fun-related activities, with paved spaces for skate boarding and relaxation areas; the second is a more naturalistic area, with trees and wildflowers, where the visitors can have a more intimate and emotional contact with nature. The areas dedicated to wildflowers allow to observe the nature and spread knowledge about the vegetation. Moreover, a view on the torrent will allow to create a relation between the park and the water flow.

我们把公园分成两个区域：第一个服务于娱乐相关的活动，里面有休闲场地，还有铺设的地面，可用来滑板；第二个更具有自然主义风格，有树和银莲花，在那儿游客可以和大自然进行更亲密的情感接触。长着银莲花的地方利于人们观察自然和传播植物知识。而且，小河边的景致会使公园和水流之间建立联系。

A guided path passes through the entire park. A series of Cor-Ten steel steps give movement to the park and help to overcome the problem of existing gradients. The material for the paved areas is concrete and the one for the path is gravel.

一条带指示的小路穿过整个公园。一系列耐腐蚀高强度钢做的台阶赋予公园运动感，帮助克服那些存在的斜坡带来的麻烦。地面是混凝土铺筑的，路是碎石铺的。

The park has also a hydraulic function: the most naturalistic part of it can be employed as area to accumulate water in case of heavy rains and secure the urban areas.

这个公园还具有水利功能：公园里最天然的部分在下大雨的时候能用来蓄水，给城市区域带来了安全保障。

S. Anna Park

圣安娜区公园

LOCATION：Lucca，Italy
项目地点：意大利 卢卡

GARDEN AREA：22,000 m²
花园面积：22 000平方米

COMPLETION DATE：2006
完成时间：2006 年

PHOTOGRAPHER：Gianfranco Franchi
摄影师：Gianfranco Franchi

DESIGNER/ARCHITECT：Gianfranco Franchi
设计师／建筑师：Gianfranco Franchi

S. Anna Park

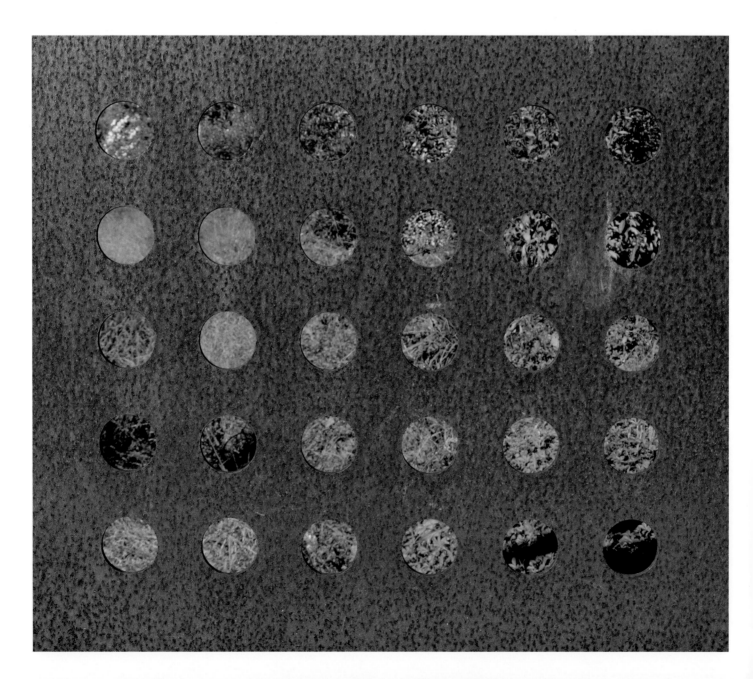

The park is found in the suburb of the city of Lucca, in the S. Anna District. The park has the function of limiting the borders of the city and constituting a filter for the surrounding countryside. It is located between a shopping centre and a residential neighbourhood.

公园位于圣安娜区卢卡市的市郊，不仅界定了城市的界限，而且还形成了附近村庄的过滤器，成为介于购物中心与住宅小区之间的一个公共空间。

The park is part of a network of parks linked to the Serchio River, which surrounds the city.

公园是与塞尔基奥河相连的公园网络的一部分，将城市围绕起来。

When we designed the park, we tried to understand how we could create a socially useful space while keeping its historical memory alive.

在对公园进行设计时，我们试图找到一种方法：既能打造一个具有社交功能的空间，又能保留其历史意义。

We wanted to characterise the new park as if it had always been there in the mind of the visitors. Then we tried to find which project and which design could be used and which materials and shapes would best suit the environment.

我们希望使新公园对于游客来说看起来像是一直存在的一样，于是我们尝试着寻找可以借鉴的项目和可以使用的方案，以及最适合这个环境的材料和造型。

We decided to use a geometric mesh and poor materials such as concrete and Cor-Ten steel, considering other shapes were not appropriate for what we wanted to do.

鉴于其他造型不适合我们所希望打造的环境，我们决定采用几何网状的造型及混凝土和耐腐蚀高强度钢等材料。

The project redevelops a part of the countryside and the use that has been done of the trees recalls the typical cultivations of the surrounding environment.

本项目重新开发了乡村的一部分，而对于树木的利用则重新恢复了对周围环境的典型塑造。

The park is characterised by small squares, artificial hills of geometric shape, and various playgrounds and spaces where people can relax and rest.

小广场、具有几何造型的假山和各种运动场以及放松和休息空间成为公园的特色。

A path which passes through the entire park, from east to west, is characterised by metallic structures where, during spring time, different types of plants flourish magnificently.

一条小路从东向西贯穿整个公园，小路旁边设置了一些金属构筑物，为小路增加了特色。春天的时候，各种植物在这里竞相生长。

Big metallic vases recall the traditional use of clay pots in the countryside.

大金属花瓶不禁使人们想起以前乡村地区经常使用到的粘土烧制的花盆。

The structure of the vegetation is characterised by Populus Alba and Populus Nigra; typical cultivations of the environment.

公园主要采用白杨和黑杨进行绿化，这两种植物是打造空间环境的典型植物。

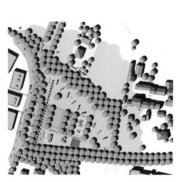

Square in Castellina Marittima

Castellina Marittima 广场

LOCATION: Italy
项目地点: 意大利

AREA: 14,000 m²
面积: 14 000 平方米

DESIGNER/ARCHITECT: Gianfranco Franchi
设计师 / 建筑师: Gianfranco Franchi

Square in Castellina Marittima

Castellina Marittima is a small village situated on the hills of the Mediterranean coast, and surrounded by woods and olive trees.

Castellina Marittima 是一个小村落，坐落于地中海沿岸的山上，四周环绕着树林和橄榄树。

The new square, even if very small (1,400 m²), aims at representing a new central point for the population of Castellina Marittima. We believe that the size of the square is balanced to the dimensions of the small village and with time it will become a new central meeting point for the population.

这个新广场，虽然很小（1 400平方米），但致力于成为 Castellina Marittima 村居民一个新的聚焦点。我们相信，由于广场的大小与这个小村庄的面积相协调，随着时间的推移，它会成为村民们新的聚会中心。

The square is placed above a car—park and therefore the possibility to include vegetation is very limited.
The spaces in the "square garden" are organised in a way that they facilitate social functions as well as ordinary outdoor and relaxation activities and local public events.

这个广场置于一个停车场的上方，因此种植植物的可能性十分有限。
广场花园里的那些空间区安排有序，目的是为社交功能、大众的户外及休闲活动和当地的公共活动提供便利。

The square is characterised by its balanced correlation of green areas and paved spaces. The two "bowers-grids" which delimit the square, both from Via Della Repubblica and the car-park situated below, are an successful way to expand the green areas. This method allows to close views and develop important perspectives, which reach the Tyrrhenian Coast.

The square is divided into two parts: the first strictly used as playground for children and young people and the second used as gathering point and outdoor space. These two spaces are delimited by a row of trees that function as filter between these different types of activities.

广场的特色在于它的绿色区域和铺筑空间搭配得很和谐。两个凉亭用的栏栅是广场的边界，把Via Della Repubblica和下方的停车场分隔开，并且很好地扩展了绿色空间。这种方法有助于把景致关在园内，同时带来了非同寻常的远景，可看见伊特鲁里亚海岸。

广场分为两部分：一部分完全是孩子们和年轻人的游乐场，另外一部分是聚会和户外空间。一排树位于这两个空间之间，起到了把不同种类的活动分隔开的作用。

The materials employed are: coloured concrete, wooden planks and steel (for the bowers).

This new space becomes square the and garden at the same time, integrating itself with the buildings around and with the green spaces nearby.

使用的材料有：有色混凝土，木板和钢筋（用于凉亭）。

这块空间既是广场又是花园，和周围的建筑及附近的绿色空间融为一体。

Jane Hansen

Jane Hansen was born in Yakima, Washington in 1963. She received her landscape architecture degree from the University of Oregon and then moved to San Francisco, California, where she worked with Angela Danadjieva and then Peter Walker. Upon returning to Portland, she worked with Walker Macy prior to founding Lango Hansen Landscape Architects in 2000 with her partner, Kurt Lango. Jane has served on a number of design advisory panels, including the Portland Design Commission.

Kurt Lango

Kurt Lango was born in Los Angeles, California, in 1963. He is a landscape architecture graduate of the University of Oregon and began his career working with Hargreaves Associates in San Francisco. In 2000, he formed Lango Hansen Landscape Architects with his Partner, Jane Hansen. He is active in the arts community having served on a number of local art juries and has lectured about the firm's work throughout the northwest.

Lango Hansen Landscape Architects

Lango Hansen Landscape Architects

It provides a wide range of services in landscape architecture, planning, and urban design. Over the past ten years, the firm's principals have successfully designed public parks, urban plazas, school and university campuses, corporate headquarters, private residences, and public facilities. Lango Hansen approaches each project as a unique opportunity to develop designs that address the particular character of the site, the specifics of the program, and the needs of individuals and communities. Using a variety of media, such as models, sketches, and computer-aided tools, the firm explores integrated design solutions. With a commitment to detail and craftsmanship, Lango Hansen creates long-lasting designs that express the innate character and value of each landscape.

Tenth Street Green Street

绿色 "第十街"

LOCATION：Lake Oswego，USA
项目地点：美国 奥斯威戈湖

COMPLETION DATE：2008
完成时间：2008 年

DESIGNER：Lango Hansen Landscape Architects
设计师：Lango Hansen Landscape Architects

DESIGN COMPANY：Lango Hansen Landscape Architects
设计公司：Lango Hansen Landscape Architects

Tenth Street
Green Street

With a right of way that measures 100 feet wide for most of the project area, Tenth Street offered a unique opportunity to help satisfy the City's larger sustainability goals and work with residents to create "green" environment in the front of their residences. In 2006, the City's Design Consultants led a series of public workshops, one—on—one meetings with residents, and distributed neighborhood newsletters to develop a master plan for the green street project. As many of the homeowners had "acquired" and planted within the 100—foot public right of way, the designers spent a significant amount of time working with individual homeowners to address individual needs for each property owner and to satisfy the City's overall sustainability goals. After a successful Master Plan process, the construction began in 2007, and completed in early 2008.

"第十街"公路用地宽100英尺，占用了大部分项目用地。"第十街"可以帮助城市实现更大的可持续性目标，并与居民一起在家门前打造"绿色"环境。2006年，城市设计顾问组织了一系列公共讨论会，与居民协商，并对区域的通信进行分布，制定出绿色街道项目的总体规划。由于很多房主都"占有"了这片100英尺的公共公路用地，并在上面种植了一些植物，因此设计师花费了大量的时间与房主沟通，试图满足每个房主的要求，同时还可以达到城市总的可持续性目标。在成功地制定出总体规划之后，项目于2007年开工，2008年初竣工。

The green street works as an interconnected drainage and water quality system for over 2,500 linear feet along the City's six blocks. A series of culverts under driveways, roadbeds and between water quality facilities move the water from garden to garden, filtering the surface water from contaminants and toxins before entering the City's storm water system. These rain gardens are planted with native and drought tolerant plants that vary in texture, color and form. The planting concept creates linear ribbons of complimentary plant species that move through the water gardens for the extent of the project while dipping in and out of the residential properties to create a cohesive planting design. Wherever possible, the water quality gardens blend with existing residential gardens to blur the boundary between public space and private yard.

绿色街道成为城市六个街区（长度超过2 500英尺）的综合给排水质量系统。车道和路基下面以及水质设施之间的涵洞把水从一个花园引向另一个花园，在水进入城市雨水系统之前将其中的污染物和毒素过滤掉。这些雨水花园中种植了质地、颜色和形状各异的本地抗旱植物。种植理念是通过欢迎状的物种打造出一条条彩带，贯穿整个项目的水景园，在住宅中穿梭，相互融合在一起。在可以实现的情况下，水景园与原有的住宅花园融合在一起，模糊了公共空间与私家花园之间的界限。

Winding through the planting is a six-foot wide walkway that connects to the existing homeowner entries and the limited on-street parking. The walkway provides an opportunity for homeowners and neighbors to meet and chat along the gardens and provides important connections to downtown and a neighboring school. The three main entries into the green street are marked with stone water gardens. These gateways are meant to provide a threshold for motorists and pedestrians entering the green street and also perform storm water quality functions.

一条弯曲的6英尺宽的人行道将原有房屋的入口和狭窄的街道停车场联系在一起。人行道使房主和邻里可以沿着花园边聚在一起闲聊，提供了一条与市区和附近的学校相连的重要通道。水景石园成为进入绿色街道的三个主入口的标识。这些人行道意在为乘客和行人进入绿色街道提供一个入口，同时也起到了改善雨水质量的作用。

Block

街区

LOCATION：Portland，USA
项目地点：美国 波特兰

DESIGNER：Lango Hansen Landscape Architects
设计师：Lango Hansen Landscape Architects

DESIGN COMPANY：Lango Hansen Landscape Architects
设计公司：Lango Hansen Landscape Architects

Block

LOCATION：Portland，USA
项目地点：美国 波特兰

DESIGNER：Lango Hansen Landscape Architects
设计师：Lango Hansen Landscape Architects

DESIGN COMPANY：Lango Hansen Landscape Architects
设计公司：Lango Hansen Landscape Architects

The Block 47 temporary landscape transforms a vacant parking lot into a lush, urban garden. The design concept stems from the site's historic tax lot map, which spanning over 100 years, is compiled into a single frame to create a mosaic of plantings, performance spaces, and seating areas.

47号街区的临时景观把一片用于停车的空地打造成了一个草木茂盛的城市花园。设计理念来自这里的历史税区地图，在跨越了100年以后，这个税区地图演变成了一幅由植物、表演空间和座位区组成的组合图画。

Using native grasses, recycled materials from demolished local buildings, gravel, and stone, the garden fabric juxtaposes color, texture, and materials. In the center, a stone mound punctuates one of the historic gravel plats and recalls the debris piles once found on the site. The site slopes 8 feet from north to south and accentuates the layers of the different garden rooms. Garden plantings were chosen to offer variety and seasonal interest throughout the year. Over time, as the grasses spread into adjacent plots, the garden fabric will evolve and break down the history—based geometry. After five to ten years, the garden will be demolished to make room for a future building.

花园采用本地草类、从拆毁的当地建筑中回收的材料、砾石和石材，将色彩、肌理和材料编织在一起。在花园的中心位置有一个石堆，强调了这个地区原来的一个砾石厂，使人不禁会想到瓦砾堆。场地从北向南倾斜8英尺，突出了不同花园空间的层次感。花园选用园林植物，在一年四季中打造出变换的四季景象。随着时间的推移，草会蔓延到周围的地块，花园也会不断扩大，打破原来的地形。5到10年后，这个花园就会被拆毁，用来建设大楼。

The garden is one of the few park—like spaces within the surrounding bustling transportation corridor. Many people from nearby businesses use this garden as a place to relax and eat lunch during the day.

在周围繁忙的交通走廊中，像这样的花园空间很少。附近的很多商业人员都把这个花园当成放松的场所，有些人还会在这里吃午餐。

The Columbian Plaza

哥伦比亚广场

LOCATION: Columbia, USA
项目地点：美国 哥伦比亚

AREA: 16,187 m²
面积：16 187 平方米

AWARD: LEED Gold certification
奖项：绿色环保认证金奖证书

DESIGN COMPANY: Lango Hansen Landscape Architects
设计公司：Lango Hansen Landscape Architects

The Columbian Plaza

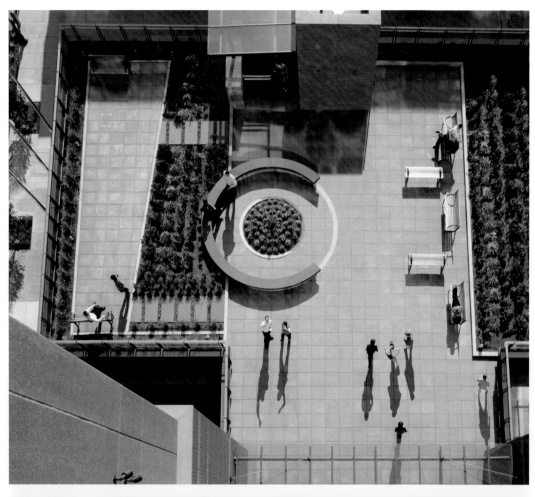

The Columbian Plaza is situated on a 4acre site adjacent to a heavily used park. For this highly visited site, Lango Hansen designed a plaza to serve as a forecourt for employees and visitors to the Columbian Newspaper. Special design focus was placed on making the plaza a comfortable, every—day gathering space that is flexible enough to allow for larger, community events throughout the year. Lango Hansen also designed a roof terrace to provide space for larger business or fund—raising events and lunch gatherings. The design used a palette of native and drought—tolerant plants, and an irrigation system that significantly reduces the site's water demand.

哥伦比亚广场位于一个4英亩的地块上，附近是一个使用频率很高的公园。由于这里的使用频率很高，Lango Hansen 为哥伦比亚报的员工和来宾建造了一个广场，作为前院。设计意在打造一个既可以用于日常聚会，又可以常年用于大型社区活动的舒适场所。Lango Hansen 还设计了一个屋顶平台，为更大的商业或集资活动和午餐聚会提供空间。设计使用了本地抗旱植物，打造出五彩缤纷的色彩，大大地减少了场地的用水需求。

Hotel Modera

摩德拉酒店

LOCATION: Portland, USA
项目地点：美国 波特兰

AREA：644 m²
面积：644 平方米

LANDSCAPE ARCHITECT：Lango Hansen Landscape Architects
景观设计师：Lango Hansen Landscape Architects

LANDSCAPE DESIGN TEAM：Jane Hansen, Kurt Lango, Elaine Kearney, Andrea Saven
景观设计团队：Jane Hansen, Kurt Lango, Elaine Kearney, Andrea Saven

ARCHITECT：Holst Architecture
建筑设计：Holst Architecture

DESIGN COMPANY：Lango Hansen Landscape Architects
设计公司：Lango Hansen Landscape Architects

Hotel Modera

The outdated motor-court of a 1960's era motel in downtown Portland has been reborn as a vibrant social space that is a distillation of the verdant Pacific Northwest landscape beyond. The reclaimed courtyard, featuring a fern-draped living wall, wood screens and crackling outdoor fire pits, is now the anchor for a restaurant and a completely renovated boutique hotel. Located on the newly redesigned transportation mall near Portland State University, the MAX light rail line slips by as hotel guests, restaurant patrons, and even curious passers-by enjoy the urban oasis.

已经过时的 20 世纪 60 年代的汽车旅馆在波特兰市区重新复兴起来，成为一个充满活力的社会空间，提取了远处碧蓝的西北太平洋景观之精华。翻造的庭院以一面爬满蕨类植物的植物墙、木屏风和噼啪作响的室外火坑为特色，成为一家餐厅和彻底翻新的精品酒店的栖息之所。酒店位于波特兰州立大学附近新设计的交通大道上，MAX 轻轨列车从这里经过，然而不论是酒店的客人、餐厅的顾客还是好奇的过路人，都沉浸在这片城市绿洲之中，丝毫不会注意到列车的经过。

Sculptures crafted from recycled granite mark the transition from streetscape to courtyard (really a roof deck over the parking garage below), where a sheltered meranti wood boardwalk leads to the transparent hotel lobby. To the left of the boardwalk, earthy decomposed granite paving, weathered steel fire pits with orange glass cullet pieces, and oversize planters define intimate seating areas for the restaurant's outdoor dining and lounge area.

回收的花岗岩雕刻而成的雕塑标志着街道与庭院之间的过渡（实际上是一个屋顶平台，下面是停车场），在这里，一条有顶的柳桉木人行道通往酒店通透的大厅。人行道的左侧有一条用风化的花岗岩铺成的小路，一个风化钢打造的火坑，里面有桔黄色的玻璃碎片，和一个超大的花盆，衬托出亲切的座位区，可以作为餐厅的室外餐饮或休息区。

A 64 feet long living wall is the focal point of the lush garden on the right side of the boardwalk. Ferns, moss, evergreen shrubs and grass create an effect of the garden continuing up and over the courtyard wall to the green roof above. The living wall is a pixilated abstraction of the Pacific Northwest itself, and weathered steel panels allude to outcrop of basalt set within the rich textures of the forest.

人行道右侧 64 英尺的植物墙是花园的亮点。蕨类、藓类、常绿灌木和玻璃打造出花园不断上升、越过庭院墙，一直延伸到绿色屋顶的效果。植物墙是西北太平洋的一个抽象元素，风化的钢板暗喻着露在森林丰富的肌理外面的玄武岩。

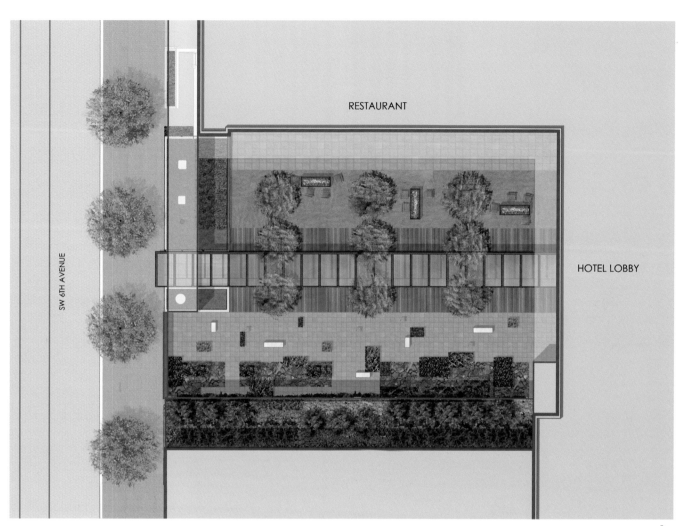

RESTAURANT

SW 6TH AVENUE

HOTEL LOBBY

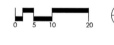

0 5 10 20

N

Jon Storm Park

Jon Storm 公园

LOCATION：Oregon, USA
项目地点：美国 俄勒冈

DESIGNER：Lango Hansen Landscape Architects
设计师：Lango Hansen Landscape Architects

DESIGN COMPANY：Lango Hansen Landscape Architects
设计公司：Lango Hansen Landscape Architects

Jon Storm Park

The design of Jon Storm Park provides opportunities for dramatic views of the Willamette River, lawn areas for flexible use, and trails that wind through riparian habitat. The park provides a number of walkways that provide direct connections to the parking lot, restroom, transient walkway and a cantilevered stand. The cantilevered stand dramatically overhangs the existing sheet pile wall and offers views of the Willamette Falls and the river below. Adjacent to the main walk, interpretive signage relays the history of the area and the importance of this open space to the development of Oregon City.

在 Jon Storm 公园可以欣赏到 Willamette 河的壮丽景观。公园里设有草坪区，可以用于各种用途；小径在河边蜿蜒伸展。公园里有许多人行道，可以直接通往停车场、休息室、过渡人行道和一个悬空的看台。悬空看台从板桩墙伸出，在上面可以欣赏 Willamette 瀑布及下面河流的壮观景色。主通道附近的标识牌传递着这个地区的历史气息，同时也反映出这片空地对俄勒冈市发展的重要性。

The parking for the park is located on ODOT property underneath the I-205 overpass. 30 parking spaces are provided along with bioswales planted with native grasses that filter drainage off the asphalt surface before it enters the river. ADA spaces are provided in the parking area and all walks within the park meet accessibility requirements.

公园的停车区安排在俄勒冈州运输部（ODOT）的地产上，位于 I-205 立交桥的下面。30 个停车位沿着种满本地草类的生态沼泽地布置，生态沼泽地可以在沥青路面排出的污水进入河流之前将其过滤。停车场内设置了符合美国残疾人法案（ADA）的停车位，公园内所有的通道都符合方便进出的要求。

In addition to picnic tables within the grass areas and adjacent to the walkway, a number of stone walls have been constructed which provide seating areas and also recall the historic stone walls that are found throughout Oregon City. These walls provide areas for people to gather and enjoy the view, eat lunch and watch the riverboats along the river. A picnic shelter is also located near the stand.

除了草坪区和人行道附近的野餐餐桌以外，公园内还设置了许多石墙，可以用作座位区，使人联想到了俄勒冈市随处可见的历史石墙。有了这些石墙，人们就可以聚在一起欣赏风景，吃午餐，或欣赏河边的河船。看台附近还有一个野餐屋。

As part of the park improvements, the City repaved Clackamette Drive which is the major street adjacent to the park and constructed a large turnaround across from the park. This was done for the City's trolley and tour buses which make their way down this street that has no outlet.

作为公园改造的一部分，俄勒冈市重新铺设了公园附近的主干道 Clackamette 车道，并且在公园对面修建了一条宽敞的回车道。这条街道本来没有出口，修建回车道以后，城市里的电车和旅行车就可以从这里通过。

The improvements at Jon Storm also included a trail called the Willamette Trail which connects Jon Storm Park to the existing Clackamette Park. This multi model asphalt trail is 12-foot-wide and winds through areas that were replanted with native vegetation to satisfy various requirements.

Jon Storm 公园的改造还包括一条名为"Willamette 小径"的通道，这条小径把 Jon Storm 公园和原来的 Clackamette 公园连接在一起。这条多模式的柏油小路宽 12 英尺，在种满本地植物的地面上蜿蜒伸展，以满足各种要求。

Story Mill Neighborhood

面粉厂的故事

LOCATION：Montana, USA
项目地点：美国 蒙大拿

AREA：56,656 m²
面积：56 656 平方米

DESIGNERS：Lango Hansen Landscape Architects，GBD Architects
设计师：Lango Hansen Landscape Architects，GBD Architects

DESIGN COMPANY：Lango Hansen Landscape Architects
设计公司：Lango Hansen Landscape Architects

Story Mill
Neighborhood

The design for the Story Mill Neighborhood connects LEED standards for green building and sustainable landscape design with historic preservation to create a diverse setting for a healthy, thriving community. The site is located around building remnants from a historic flour mill community at the base of Baldy Mountain near downtown Bozeman, Montana. The neighborhood design includes the preservation and reuse of historic building remnants, new residential and multi-use development, 56,656 m² of new parks, a vast network of pedestrian and bicycle trails, ecoroofs, stormwater gardens, and ecologically sensitive habitat preservation. Focus was placed on balancing dense living and working spaces with a rich network of parks and public outdoor spaces to fulfill needs for various recreational activities, quiet contemplation, large gatherings, community gardens, wildlife corridors, and exposure to nature.

"面粉厂的故事"的设计将绿色建筑的LEED标准、可持续景观设计与历史保护相结合，为这个健康而又充满活力的社区打造出多样化的环境。场地位于蒙大拿州波兹曼市区附近的秃山脚下，这里原来是一个面粉厂，现在还留有一些建筑遗迹。其设计包括对建筑遗迹的保留与再利用、新住宅与多用途建筑的开发、56 656 平方米的新公园、庞大的人行道和自行车道网络体系、生态屋顶、雨水花园，以及敏感生态栖息地的保护。设计强调多样化的生活和工作空间与丰富的公园和公共室外空间网络之间的平衡，以满足各种娱乐活动、静思、大型聚会、社区花园、野生动植物走廊和亲近大自然的需求。

Careful consideration was given to protecting ecological resources and meeting the needs of a dense and diverse community. As part of the project, design guidelines were written and put into all new development within the 364,217 m² of the Story Mill Neighborhood. These guidelines include preserving and enhancing the site's natural features, protecting views, highlighting cultural resources and respecting the site's existing topography. Design guidelines encourage bioswales, stormwater planters, flow spreaders, pervious paving, disconnected downspouts and/or rain chains, and green roofs. Guidelines for outdoor public spaces include the integration of stormwater features, where possible, to celebrate water collection and provide opportunities for public awareness and active participation. All stormwater is to stay within each lot by being either filtered or reused. Landscape design guidelines were written to require up to 50% of the privately developed open space to be planted with native species.

为了保护生态资源、满足多样化密集社区的要求，设计师进行了周密的考虑。作为项目的一部分，相关人员制定了设计指导原则，并将其应用到364 217平方米"面粉厂的故事"小区新开发的项目中。项目指导原则包括保护并加强场地的自然特性、保护景观、突出文化资源和尊重场地原来的地形。指导原则鼓励采用生态沼泽、雨水灌溉的培植器皿、水流分布器、透水透气路面、分离下水管和／或雨链，及绿色屋顶。户外公共空间的设计指导原则是将雨水特性融合在一起，在可能的情况下，可以鼓励使用雨水收集系统，提高公众节水意识，鼓励公众积极参与其中。每个地块中的所有雨水都被收集起来，进行过滤或者重新利用。景观设计指导原则要求50%的私人开发商开发的户外空间都应种植本地物种。

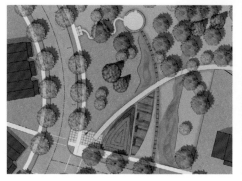

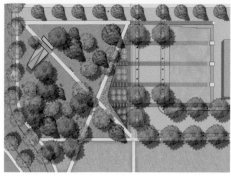

Shlomo Aronson Architects

Office awards

2010 Shlomo Aronson honored as one of the 101 most positively influential people in Israel for the year 2009/2010, The Marker Magazine

2010 Prize of Honor, Domus Magazine, Israel, for Herzeliya Park

2009 Projects of the Year, Architecture of Israel Magazine, for Herzeliya Park

2005 Honor Award, American Society of Landscape Architects, for Ben Gurion International Airport

2004 The Olmsted Lecture Harvard University

2001 Distinguished Alumnus Award, University of California, Berkeley

2000 Jerusalem Prize for Architecture

1999 Silver Medal for Design, Kunming EXPO, China

1998 Israel Architects and Town Planners Award Prize

1997 Karavan Prize, for Kreitman Plaza

1996 Represented Israel in the International Biennale in Venice, for Shaar Hagai interchange, Nazareth, Cross Israel highway.

1995 "Designer of the Year" (with Arch. David Resnik) for the magazine "Mivnim" for the Master Plan of the town of Beit Shemesh.

1995 Excellence in Communication, Landscape Architectural Magazine

1991 Represented Israel in the International Biennale in Venice with the Sherover Promenade

1991 "Beautiful Israel" Prize, for the Sherover Promenade

1990 Rechter Prize, for the Susan Delal Dance and Theater Center

1990 The Gold Medal, and Best Design Award, Osaka EXPO

1989 Pfefferman Prize, for the Beit Shalom Park

The projects are designed and executed in our office by an experienced and skilled team of architects and landscape architects, headed jointly by landscape architect Barbara Aronson and architect Ittai Aronson, advised by landscape architect and office founder Shlomo Aronson.

Partners: Shlomo Aronson (right), Barbara Aronson (left), Ittai Aronson (middle)

Shlomo Aronson Architects

The firm of Shlomo Aronson Architects was founded four decades ago. Throughout these years we have designed and developed hundreds of projects, mainly in Israel but also abroad. Over the years our multi-disciplinary office has acquired a varied and rich expertise in different fields of architecture and landscape architecture, from national, regional and local master plans to the detailed design of landscape architectural projects, architecture and project supervision. We believe in practicing architecture and landscape architecture jointly on the widest platform possible: it is much more productive to design a landscape or a building complex as part of a comprehensive design philosophy that you help to formulate at the scale of policy—making. To this day the specialty of our office is not to specialize on a particular aspect of our profession but to plan projects from their conception, from the master plan phase, to their construction. Good planning influences good architectural design, and vice versa.

American Consulate

美国领事馆

LOCATION: Jerusalem, Israel
项目地点: 以色列 耶路撒冷

COMPLETED DATE: 2010
完成时间: 2010 年

LANDSCAPE ARCHITECT: Barbara Aronson
景观设计师: Barbara Aronson

ARCHITECT: Mann Shinar Architects and Planners
建筑师: Mann Shinar Architects and Planners

DESIGN COMPANY: Shlomo Aronson Architects
设计公司: Shlomo Aronson Architects

American Consulate

The landscaping of the American Consulate in Jerusalem includes the design of a series of open spaces inside and outside the compound: entrance plaza, a courtyard for visitors to the consular section, parking areas, a green roof and the garden areas around the main building. The design intent is to develop a common language for all of these open spaces that reflects at the same time the character of surroundings.

The formal layout of all outdoor spaces is offset by delicate variations in the treatment of the details: local stone paving and cladding is used to create different patterns by varied combinations of stone sizes and finishes. Different types of trees and plants create subtle differences in the feel of the various open spaces.

For the first stage of the project, the main building was implemented only partially.

驻耶路撒冷的美国领事馆,其环境美化包括了一系列内部和外部的复合式开放空间设计:入口广场、停车场、绿色屋顶、供访客休息的领事部庭院和主体建筑周围的园林区域。其设计意图是为所有这些开放性的空间制定一种共通的理念,并且与此同时能够反映周围环境的特点。

所有户外空间的形式布局,可以通过建造细节中的微妙雅致的变化来完善:运用当地石材的铺砌和包层创造出了各种不同的形态,比如由石头的大小和罩面漆的变换运用来形成多姿多彩的搭配组合。不同种类的林木和花草植物创造出了各种开放空间感的细微差异。

该项目的第一期主要对主体建筑的局部位置进行施工。

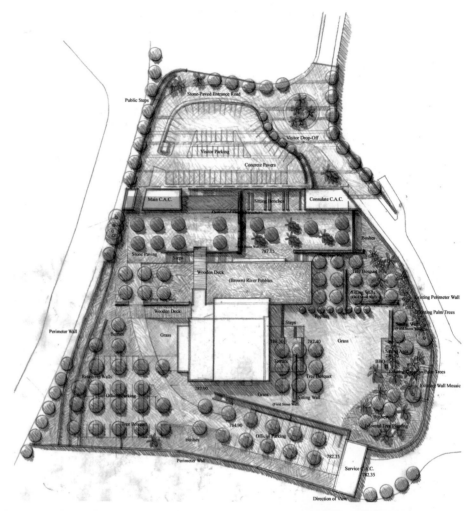

Public Steps

Stone-Paved Entrance Road

Visitor Drop-Off

Visitor Parking

Concrete Pavers

Main C.A.C.

Sitting Benches

Consulate C.A.C.

Bushes

Stone Paving

Steps

787.1s

Wooden Deck

(Brown) River Pebbles

Tree Bosquet

Sitting Walls

Existing Perimeter Wall

Wooden Deck

Poplars

Existing Palm Trees

Perimeter Wall

Bushes

Grass

Steps

Grass

Sitting Walls

Retaining Walls

784.30

787.40

BBQ

Existing Georgian Palm Trees

Official Parking

Terrace

Tree Bosquet

Existing Wall Mosaic

782.90

Sitting Wall

Tree Bosquet

Grass

Bushes

Tree Bosquet

784.90

Bushes

Official Parking

Informal Tree Planting

Perimeter Wall

782.35

Service C.A.C.

782.35

Direction of View

Har Adar # 1

Har Adar #1 住宅

LOCATION：Har Adar，Israel
项目地点：以色列 哈尔阿达尔

COMPLETION DATE：2009
完成时间：2009 年

SITE AREA：500 m²
占地面积：500 平方米

TEAM：Ittai Aronson，Tal Bilinsky
团队：Ittai Aronson，Tal Bilinsky

ARCHITECTS：Ittai Aronson in partnership with Tova Dagan
建筑师：Ittai Aronson 与 Tova Dagan

DESIGN COMPANY：Shlomo Aronson Architects
设计公司：Shlomo Aronson Architects

Har Adar # 1

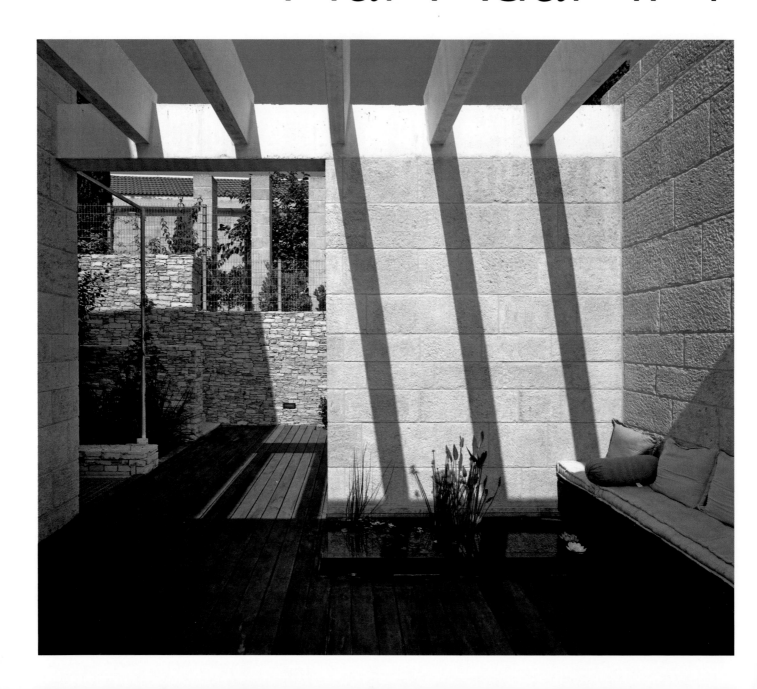

This house on a long, narrow and steep lot is centered around a big, sheltered courtyard with a mulberry tree in its middle. A double height space faces this outdoor patio and is revealed as a surprise to people entering the house. This arrangement blends the indoor and outdoor seamlessly, providing the family with a private garden all year around. The material palette for the walls and floors was intentionally limited to one type of natural stone, dressed in different ways.

住宅位于一个狭长、略陡的位置，周边是一个宽敞的庭院，院落中央有一株提供阴凉的桑树。正对着户外的天井处设有一个双层高度的隔间，进入住宅的人们看到它会大感惊讶。院落的布局让室内和室外浑然天成地融合在一起，给住户提供了一个全年的私家花园。墙壁和地板采用同一种材料——一种天然岩石，虽是同一材料，但最后呈现出的形态却完全不同。

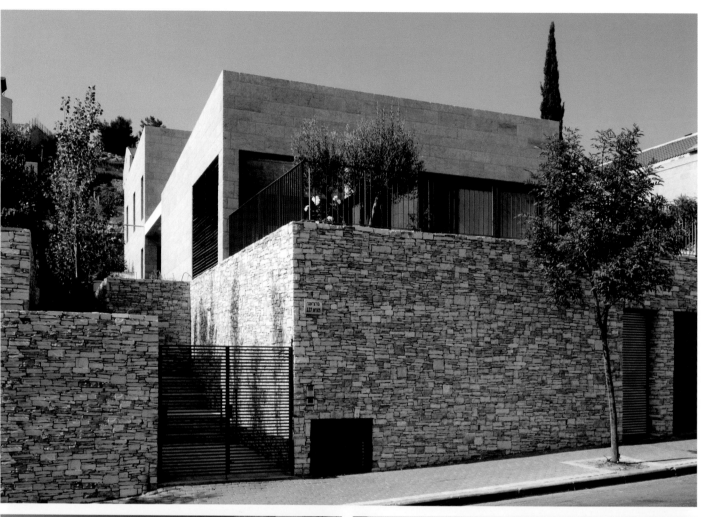

Har Adar #2

Har Adar #2 住宅

LOCATION: Har Aadar, Israel
项目地点：以色列 哈尔阿达尔

COMPLETION DATE: 2009
完成时间：2009 年

SITE AREA: 500 m²
占地面积：500 平方米

TEAM: Ittai Aronson, Tal Bilinsky
团队：Ittai Aronson, Tal Bilinsky

ARCHITECTS: Ittai Aronson in partnership with Tova Dagan
建筑师：Ittai Aronson 与 Tova Dagan

DESIGN COMPANY: Shlomo Aronson Architects
设计公司：Shlomo Aronson Architects

Har Adar #2

The design of this house which is on a long, narrow and steep lot in Har Adar has to take into account the severe winds that are typical for this neighborhood. The house is designed around two inner courtyards that are private, sheltered and shaded, with large windows facing them. All the public spaces are interconnected with the patios and the rest of the garden to create a mosaic of indoor and outdoor living. Negotiating the height differences between the house and the street creates the opportunity to provide a wide entrance leading up to the central patio.

住宅位于 Har Adar 的一个狭长且陡峭的地块上，因此对其进行设计首先要考虑的就是这一地区典型的大风气候。住宅有一个大窗户，面对着两个可以乘凉的私密内院。所有的公共空间都与中庭和花园的其他部分相互连接，将室内生活与室外活动融合在一起。设计时需要协调房屋和街道的高差，不过这反而提供了为中庭建造一个宽阔入口的机会。

Herzeliya Park

海尔兹利亚公园

LOCATION：Herzeliya, Tel Aviv Metropolitan Area, Israel
项目地点：以色列 特拉维大市区 海尔滋利亚

AREA：stage A is 161,874 m²out of 728 434 m², stage B is currently under construction
面积：A 阶段为 728 434 平方米中的 161 874 平方米，B 阶段正在建设中

COMPLETION DATE：2008
完成时间：2008 年

DESIGN COMPANY：Shlomo Aronson Architects
设计公司：Shlomo Aronson Architects

Herzeliya Park

This city park successfully combines ecological necessity with functional diversity. It provides generous recreational spaces for everyone from toddlers to retirees. The park is watered by recycled water which makes the use of much desired lawn areas sustainable. On the other hand, the contiguous winter flood basin has been preserved, supporting local and transient birdlife.

这个城市公园将生态环境的必要性与功能的多样性完美地结合起来。该城市公园为每个人，无论是小孩还是老人，都提供了足够的空间，用来进行休闲娱乐活动。一方面，这个公园通过循环水系统来浇灌草坪，使草坪生长得郁郁葱葱，并让茂盛的草坪持续生长下去。另一方面，相邻的冬季蓄洪区域已经被保护起来，这个蓄洪区被用来供养本地的鸟群和冬季迁徙来的候鸟鸟群。

The presented project is the first stage of a large future urban park (161,874 m² out of 728 434 m²). As part of the master plan that outlines a programmatic layout and development strategy for the entire site, the first stage was planned as a cornerstone. It creates an active and intensively developed park section that addresses the first wave of demands of its protential users. At the same time it protects and exposes the rich natural processes and ecologies existing on the site. It is also intended to raise public interest and political support for the protection of the bordering winter ponds and their rich wildlife, until now hidden from view and widely snubbed as mosquito breeding grounds. Adjacent to the existing soccer stadium and the city's sports center, the park becomes an integral part of the larger recreational complex of Herzeliya.

现阶段所呈现出来的项目规划是未来整个大型城市公园规划中的第一期（第一期为161 874平方米，整体规划为728 434平方米）。作为总体项目规划的一部分，一期项目规划如同基石般重要。一期项目规划将描画出一个纲领性布局的轮廓，并且为整个现场工程提供发展性的战略方案。这里解决了第一批潜在用户的需求，创建了一片生机勃勃的，并且有发展潜力的公园带。与此同时，这个城市公园保护并展现了丰富的自然进程和本地的生态环境。这也是为了提高市民保护与公园接壤的冬季蓄洪区和丰富的野生动物这一公共利益的意识，并获得更多的政治支持；直到现在为止，这片区域还是不为人所见的，仅仅被当做蚊子的滋生地，而被公众普遍忽视。公园毗邻城市的足球场和体育中心，现已成为 Herzeliya 大型综合娱乐活动场所中不可或缺的一部分。

The park site is a historical flood basin draining toward the Mediterranean Sea via an ancient Roman aqueduct (only last year a new tunnel was added). More than half of the surrounding towns' storm drainage ends up in the park, channeled previously in concrete-lined runnels toward the aqueduct. Most of the former agricultural land has been abandoned or used as landfill for excess earthworks. Large winter ponds with standing water during the winter months exist on site due to the heavy clay soil of the area. The seasonal flooding has protected this piece of land from housing development in the center of a very densely populated area.

公园地处一个有着久远历史的排洪区，蓄滞的洪水经由这里，再通过一个古老的罗马时期修建的渡槽而流向地中海（直至去年，这个地区才增加了一个新的泄洪管道）。每当暴雨到来时，周边有超过半数的城镇，都会通过这个由混凝土衬砌渠道而形成的河道将水排入公园，再流向渡槽。以前大部分的农业用地都已经荒废了，或成为多余土方的垃圾填埋场。由于该地区的土质是那种十分粘稠的黏性土，其沁水率很低，一般大型的池塘在冬季的几个月时间里一直持有积水。正是因为这种季节性的洪涝灾害，保护了这个位于人口稠密区的中心区域，使其未被用作房地产开发。

Main Plaza, Israel Institute of Technology (Technion)

以色列理工大学主广场

LOCATION：Haifa, Israel
项目地点：以色列 海法

AREA：16,187 m²
面积：16187 平方米

TEAM：Shlomo Aronson, Jorge Salzberg, Ariel Ginat
团队：Shlomo Aronson, Jorge Salzberg, Ariel Ginat

DESIGN COMPANY：Shlomo Aronson Architects
设计公司：Shlomo Aronson Architects

Main Plaza, Israel Institute of Technology (Technion)

A parking lot previously occupied the most central open space of the Technion, a campus of fifteen thousand students. The parking was moved, and in its place a stone—paved central plaza was established bordering a large sloping lawn. The lawn accommodates a large number of students for celebrations such as concerts and commencement. On the upper part of the lawn is a 75 m water feature, built from massive stone with water recycled from the air conditioning system of the campus. For special events, the fountain can be turned off and the linear waterfalls provide additional sitting places.

以色列理工大学拥有 15 000 名学生，它的校园中央有一块空地，空地以前被用作停车场。停车场被迁走后，这里建成了一个中心广场，广场用石头铺成，并且与一大片草坪坡地相连。在这片草坪上可举办容纳很多学生的大型庆典活动，比如音乐会和毕业典礼。草坪上有一个长达 75 米的水景瀑布设计，这个瀑布通过校园空调系统的水循环从巨石中喷出。在特殊情况下，可以关闭水循环系统，为学生们提供更多的休憩场所。

This area has become a popular place for students' relaxation activities.

这一区域现已成为学生们进行休闲娱乐活动的最佳去处。

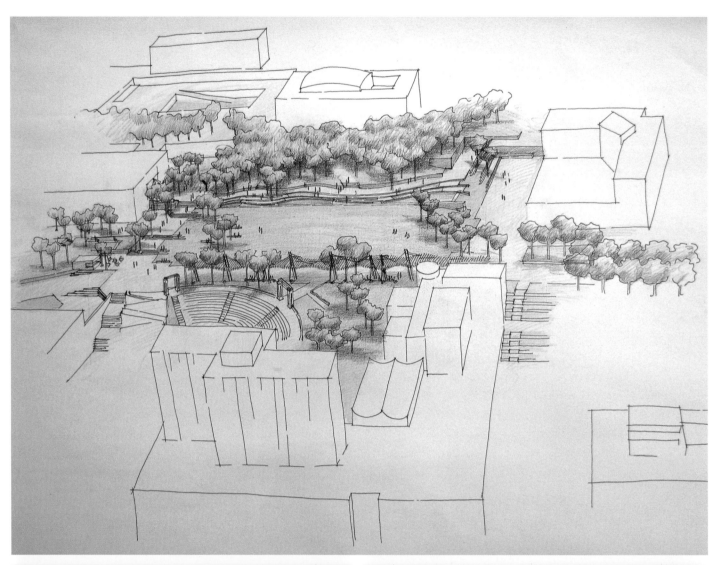

Student Union Entrance Plaza, Israel Institute of Technology (Technion)

以色列理工大学学生会入口广场

LOCATION：Haifa，Israel
项目地点：以色列 海法

COMPLETION DATE：2009
完成时间：2009 年

Team：Ittai Aronson，Jorge Salzberg，Barbara Aronson，Ofri Gerber
团队：Ittai Aronson，Jorge Salzberg，Barbara Aronson，Ofri Gerber

DESIGN COMPANY：Shlomo Aronson Architects
设计公司：Shlomo Aronson Architects

Student Union Entrance Plaza, Israel Institute of Technology (Technion)

以色列理工大学学生会入口广场

LOCATION：Haifa，Israel
项目地点：以色列 海法

Team：Ittai Aronson，Jorge Salzberg，Barbara Aronson，Ofri Gerber
团队：Ittai Aronson，Jorge Salzberg，Barbara Aronson，Ofri Gerber

At the new entrance to the student union building, the ramp for handicapped users which is often placed outside the central functions of a space, and is used here as the main generator of the design. The ramp is the "spine" of a casual communal meeting space which can also be an outdoor theater for dance performances or for informal student meetings. Wood was used to unify all design elements, turning the entire plaza into one sculptural element. The distinctive warmth of the material also provides the appropriate surface for the dancers and for the sitting audiences.

在学生会大楼的新入口有一个位于核心功能空间以外的坡道，这个坡道是专为残障人士设计的，也是设计的核心部分。这个斜坡可作为公众休闲聚会的主要场所，也可以为舞蹈表演以及非正式学生会议提供一个露天场所。所有的设计单元通过木质统一了起来，把整个广场变成一个雕塑般的整体。材料所特有的保暖性也让舞者和坐着的人感到更加舒适。

Yad Vashem Holocaust Museum Complex

耶路撒冷犹太大屠杀博物纪念馆综合建筑园区

LOCATION：Jerusalem, Israel
项目地点：以色列 耶路撒冷

DESIGN COMPANY：Shlomo Aronson Architects
设计公司：Shlomo Aronson Architects

Yad Vashem Holocaust Museum Complex

The landscape design for an institution that represents a historical event of such profound importance, from our point of view, had to be simple and unpretentious and not try to compete with the real subject matter of the institution. The design had to allow for the flow of a great number of visitors through the site in a simple and quiet way, controlling the masses of people while still allowing for private and intimate experiences for the individual visitor.

此处的景观设计是能够代表意义深远的历史史实的，从我们的角度来看，设计风格是要简约并谦逊的，而不是试图突出展现博物馆的现实题材。设计布局中要确保大批量游客能够方便、肃静地涌入博物馆；在调控群众入场的同时，仍然可以确保个体游客的亲历体验。

To this end we developed a major pathway that includes stone—paved walks and plazas linking the new museum complex with older buildings, monuments and previously developed areas. Minor paths connect to more intimate spaces and give an alternate route through the garden. In accordance with the desire for a minimalist treatment of the physical surroundings of the memorials and museum, we developed an overall language of light—washed stone—paved plazas contrasting with dense plantings of trees and bushes. Based on the same palette of materials, the individual character of each plaza was defined by different interpretations of the same design elements, by using different stone sizes and finishes, and by the use of different tree species for each space.

为此我们开发了一条主干道，含纳了用石头铺砌的步道和广场，这些步道、广场连接了新建的博物馆群与旧时建筑、纪念碑以及以前的发达区域。次干道则连接了更多幽谧的空间，并且为游人游走公园提供了备用路径。我们研发了一种用易清洗的石头铺成广场的整体风格，与茂密的树林和灌木丛相映成趣，这正迎合了纪念馆和博物馆的实体环境追求简约风格的理念。基于材料相同的颜色，并且通过不同大小的石头和罩面漆的使用，以及每一个空间不同树种的使用，每片广场的个性都通过对相同设计元素的不同解读表现了出来。

The use of stone paving throughout the grounds contrasts with the smoothness of the elegant concrete finish of the new buildings, as well as tying the new museum complex to the previously built fabric of memorials and buildings. Custom—made concrete benches echo the treatment of the new architecture. Great care was taken to create subtle differences in the finish of the stone for each space, such as creating more reflective surfaces in the Remembrance Plaza, versus the highly textured and directional finish in the Entrance Plaza.

整个地面铺满了石头，与新式建筑群水泥罩面漆的优雅的光滑感相得益彰，这样也可以使新建的博物馆综合建筑群与以前建造的纪念馆建筑群的外立面相互搭配。所定制的混凝土长椅则呼应了新式建筑风格的设计理念。在设计中，我们精心地创建出了每处石头罩面漆的细微差异，比如在纪念广场设置更多的反射面，在入口广场铺设高质感的、极具特色的并有方向指示性的修饰面。

At a large scale, the landscaping of the museum grounds blends into its natural surroundings through the use of local forest vegetation and indigenous bushes and groundcovers. At the smaller scale, the use of a restrained selection of bushes and trees, without spectacular flowers or foliage coloration, reinforces the meditative nature of the museum and memorial sites. These almost monochromatic plantings create the background for the different plazas. Magnificent views into the surrounding Judean Hills are opened in many points in the site.

大规模的博物馆庭院的景观美化，运用了当地的森林植被、原生灌木和花卉地被等元素，实现了与其周边自然环境相互融合的完美效果。较小规模的美化，使用特定的没有绚烂花朵或显色枝叶的灌木和树木，以加强博物馆和纪念馆区域原有的容易让人沉思的氛围效果。由此可见，这些近乎单色系的景观区域，为不同的广场创建了相应的背景布局。此外，这里呈现了多处呈现了周围朱迪亚山的壮丽景象。

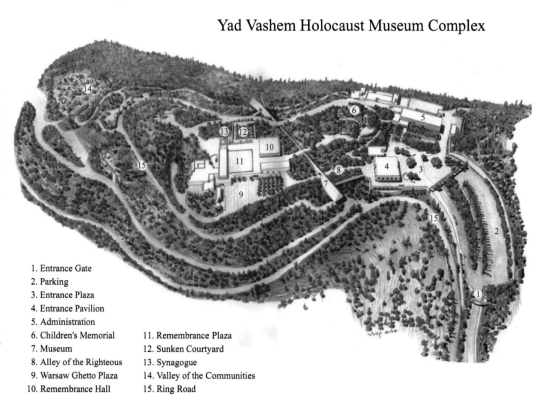

Yad Vashem Holocaust Museum Complex

1. Entrance Gate
2. Parking
3. Entrance Plaza
4. Entrance Pavilion
5. Administration
6. Children's Memorial
7. Museum
8. Alley of the Righteous
9. Warsaw Ghetto Plaza
10. Remembrance Hall
11. Remembrance Plaza
12. Sunken Courtyard
13. Synagogue
14. Valley of the Communities
15. Ring Road

Stig L. Andersson (b. 1957)

Professor, Landscape Architect Maa., Mdl.
Owner, Founding Partner (1994) and Creative Director of SLA A/S.
Honours and Memberships
Member and Head of Jury for the "Design your city" (Design din by) national design
competition for youngsters aged 15—24, 2007
Appointed Editorial Board Member of the European Foundation for Landscape
Architecture's yearbook, 2005
Member and Head of Nykredit's Jury for the Architectural Award (Nykredits
Arkitekturpris), 2000—2004

COMPANY PROFILE

SLA developes urban spaces in all scales, from master plans to installations, in Denmark and elsewhere. SLA explores innovative methods of design and research, and has developed distinctive approaches on themes such as sustainability, participatory planning and innovative preservation.

ORGANISATION

The office was founded by architect Stig L. Andersson in 1994 and has won numerous competitions on landscape and urban design in Denmark and abroad. Stig L. Andersson is owner and Creative Director, Lene Dammand Lund is the Managing Director and Hanne Bruun Møller Associate Partner.

PROFESSIONAL REGISTRATION

SLA is a member of Danish Association of Architectural firms (Danske Ark) and our landscape consultancy services are provided in accordance with the principles stated in its rules and regulations.

BUSINESS INSURRANCE

Consultancy liability insurance is with: TRYG insurance, Klausdalsbrovej 601, 2750 Ballerup, Denmark; Policy no.: 670-4.532.064.963; Coverage sum: DKK 12.5 m per insurance year: DKK 5m for personal injury and DKK 2.5m for other insurance events. Insurance is valid up to 5 years after closure of the firm. Geographical coverage: Europe.

DIGITAL TOOLS

SLA processes a fully built digital network for all architects. We use digital design projects such as AutoCAD, Land4 with layer and colour control after IBB and Quicksurf for terrain levelling and ground calculations. Microsoft Office is used for texts, for databases File-Maker Developer, besides FotoStation Pro and Adobe for presentations and visualizations. We use SketchUp for 3D modelling.

GUARANTEE OF QUALITY

SLA's guarantee of quality is in accordance with the Quality Guarantee Handbook for SLA, which is continually revised for each project. The office's own internal guarantee of quality follows BPS's formula.

COLLABORATORS

Arup, BDP Architects, Cowi Engineering, Danish Institute of Town Planning, Cenergia, Copenhagen City, Dorte Mandrup Architects, Harvard University, writer Jan Sonnergaard, Foster & Partners, Frederiksberg Municipality, Galerie Aedes in Berlin, Gehl Architects, Hansen & Henneberg, Hausenberg, Henning Larsen Architects, 2+1, artist Morten Stræde, Mette Sandbye, Lemming & Eriksson Engineers, Oluf Jørgensen Engineers, RKD Architects, Schmidt Hammer Lassen, Tetraplan, Transform, University of Copenhagen and many others.

Charlotte Garden

夏洛特花园

LOCATION: Copenhagen, Denmark
项目地点：丹麦 哥本哈根

AREA: 13,000 m²
面积：13 000 平方米

TEAM: Stig L. Andersson, Hanne Bruun Møller, Lars Nybye Sørensen
团队：Stig L. Andersson, Hanne Bruun Møller, Lars Nybye Sørensen

COLLABORATOR: Lundgaard & Tranberg
合作者：Lundgaard & Tranberg

AWARD: 2004 Nominated for the Mies van der Rohe
奖项：2004 年 Mies van der Rohe 提名奖

DESIGN COMPANY: SLA
设计公司：SLA

Charlotte Garden

Charlotte Garden is used more and more by local people; on the way to the café; as a meeting place; chatting at the playground because they live in the housing blocks of the same name that surround the open park and courtyards on sterbro in Copenhagen. The garden has become a place and a room.

越来越多当地的民众喜欢选择夏洛特花园作为休闲度假的理想场所。人们可以把这里作为去咖啡厅前碰面的地方，还可以在操场上聊天。这或许是因为大家住在哥本哈根 sterbro 的开放式公园和庭院附近的同名住宅区域。花园仿佛已经成为了一间公屋。

Viewed from the apartments above, the Charlotte Garden is a picture, constantly changing its characters in sync with the seasons, times of day and changing wind directions.

从高层公寓往下看，夏洛特花园是一幅图画，其特点会随着季节、一天的时间和风向的变化而同步更迭。

CALENDAR OF PLANTING

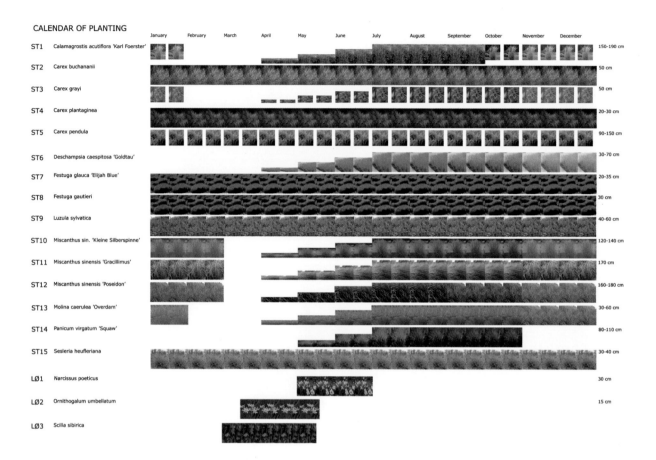

		January	February	March	April	May	June	July	August	September	October	November	December	
ST1	Calamagrostis acutiflora 'Karl Foerster'													150-190 cm
ST2	Carex buchananii													50 cm
ST3	Carex grayi													50 cm
ST4	Carex plantaginea													20-30 cm
ST5	Carex pendula													90-150 cm
ST6	Deschampsia caespitosa 'Goldtau'													30-70 cm
ST7	Festuga glauca 'Elijah Blue'													20-35 cm
ST8	Festuga gautieri													30 cm
ST9	Luzula sylvatica													40-60 cm
ST10	Miscanthus sin. 'Kleine Silberspinne'													120-140 cm
ST11	Miscanthus sinensis 'Gracillimus'													170 cm
ST12	Miscanthus sinensis 'Poseidon'													160-180 cm
ST13	Molina caerulea 'Overdam'													30-60 cm
ST14	Panicum virgatum 'Squaw'													80-110 cm
ST15	Sesleria heufleriana													30-40 cm
LØ1	Narcissus poeticus													30 cm
LØ2	Ornithogalum umbellatum													15 cm
LØ3	Scilla sibirica													

■	Perennial grasses of 1.40m - 2.00m
■	Perennial grasses of 0.70m - 1.40m
■	Perennial grasses of 0.20m - 0.70m
■	Lawn

Plan 1:500

Winter ambience in the Charlotte Garden,
together with the grasses and light forms
space for exploration.

夏洛特花园的冬季氛围与花草和光线相结
合形成了探索空间。

The Charlotte Garden is encircled by a
T—shaped residential block with a publicly
accessible building with a café and a
fitness centre. Here the Copenhagen type
of courtyard block has been opened up and
the courtyard garden has become a public
space and meeting place for the entire
neighbourhood.

夏洛特花园被一个 U 形的住宅区域环绕，
这里建有一座公共开放性的咖啡馆和健身
中心。在这里，哥本哈根式的庭院街区已
被开发建设，由此可见，庭院花园已成为
整个社区的公共区域和聚会场所。

Billowing beachscape illustrates the site's
historic relation to the sea. The blue—black
buildings encircle the soft, green pillows.

汹涌澎湃的海边景观生动地体现了该区域
与大海悠久的历史关系。深蓝色的建筑群
仿佛环绕着柔软的绿色棉垫。

The different and changing spaces are held
together by paths crossing the garden,
whilst the delineation of the spaces is
achieved by means of change of material.
The garden is a textural and sensory space
with a particular attention to nuances and
movement.

彼此不同且不断变化的区域通过穿越花园
的路径被统一到了一起。同时，通过变换
材料的方式，实现了划分不同区域的作用。
夏洛特花园是一个特别注意细微差别和变
动的富有质地和感觉的空间。

Plantings primarily consist of different grass sorts, such as indigenous grasses, blue fescue, Balkan blue grass, and purple moor grass.

园林绿化区含纳了各种不同的草种，如：本土草、蓝羊茅、巴尔干蓝草，和紫色的沼泽草等。

A conglomerate is formed by residential gardens and public spaces where small niches offer inviting places to gather.

宅第园林和很多公共场所形成了一个聚合集团，其中小型生态区提供了极具吸引力的集散地。

Unusual for the Scandinavian latitudes, there is yea-round colour in Charlotte Garden. Colours change over the course of the year, from summer nuances of blue and green to a range of golden winter hues.

夏洛特花园地处斯堪的纳维亚纬度地区，全年的颜色变化是独具特色的。一年中，颜色在不断地变化着，从夏天微妙的蓝色、绿色变化到冬季的金色色调。

The winding paths turn a walk into a spatial experience that changes with the season, and each visit offers a new experience.

蜿蜒的小径让散步成了一种空间体验，随着季节的更迭，使每次游历都成为一种全新的体验。

CLOUD

云

LOCATION: Copenhagen, Denmark
项目地点: 丹麦 哥本哈根

AREA: 5,500 m²
面积: 5 500 平方米

COMPLETION DATE: 2011
完成时间: 2011 年

TEAM: Stig L. Andersson, Hanne Bruun Møller, Signe Hertzum, Malin Blomquist
团队: Stig L. Andersson, Hanne Bruun Møller, Signe Hertzum, Malin Blomquist

COLLABORATOR: Schmidt Hammer Lassen
合作者: Schmidt Hammer Lassen

DESIGN COMPANY: SLA
设计公司: SLA

CLOUD

Copenhagen lies by the water. In recent years it has become more attractive to live and work by the water than it was, when the harbor was the city's backside with industry and heavy transportation. Many new and local neighborhoods have emerged after the industry and storage areas have been dismantled and removed. But water pollution is also a growing problem in Copenhagen: rain water. Climate changes have increased pressure on the inadequate sewerage in the city. When it rains, the sewers are threatened by overload and the polluted sewage ends up in the harbor. Water is both an amenity to the city and to people's health. It is important for us to take care of the water's qualities through an understanding of its harmful as well as its life—giving and sensual properties.

哥本哈根是依存于水资源利用与开发的城市。以前由于重型工业运输污染，海港地区成为城市生活环境的落后区，而近些年来，这里相比以前吸引了越来越多的人们到此工作和生活。很多新来的和本地的居民在工业区、仓储区拆除移走之后迁入到这里。但是，水质污染仍然是哥本哈根日益严重的问题：雨水。气候变化给城市本不完善的污水再生利用处理设施带来了越来越多的压力。当下雨的时候，污水排水道会有过载的危险，而且受到污染的污水也将排放到海港中。可见，水既是城市形象的美容师又是市民生活健康的医疗师，通过了解水的危害性和水赋予生命的灵性财富，我们懂得了水质对于人们生活发展的重要性。

The urban space CLOUD lies on the border between the old center of Copenhagen and the modern harbor, where the old city's apartment buildings, with faceted and rough exterior surfaces, meet the smooth and reflective business domiciles in a new spatial order: an order where the classic axial and hierarchical room—partition turns into the liquid, non—hierarchic spatial sequence.

CLOUD 城市空间位于哥本哈根的老中心区和现代海港之间的交界处，其中老城区的公寓建筑物建有雕琢切割过的外表面，呼应了商业住宅追求平滑反光效果的新式空间原则：这项原则是基于将有古典轴线、分层次的室内隔断转变成流线形、不分层的空间序列。

It is from these traits that the urban space CLOUD originates. Here, the Crystal—the new headquarters of a major Danish bank— rises like frozen water and reflects the sunlight with its smooth, sharp and jagged surface. On days with overcast weather, the building draws the gray shades in and combines with the ambient colors and expressions of the weather.

CLOUD 城市空间的开创正是基于这些特点。在这里，水晶大厦——一个丹麦重要银行的新总部——外表面好似冻结的水面，能够利用其光滑、锋利、凹凸不平的表面反射阳光。到了阴雨天天的时候，大厦仿佛笼罩了灰色的阴影，并且融合了周围环境的色彩和天气的变化。

Clouds are in their essence visible formations in the color spectrum from white to gray to black. They consist of small water droplets and ice crystals. Clouds are formed in the atmosphere as a result of condensation of water vapor. This can occur by cooling, addition of water vapor, a mixture of air with different temperatures or a combination of these events. The clouds' appearance and size are determined by temperature and stability conditions, humidity and wind.

云朵在其本质上有可见性的形态，有其从白色到灰色再到黑色的色彩光谱。它们是由小水滴和冰结晶构成的。不同温度的空气混合，通过冷却，可以增加空气中水蒸气的含量。在大气中的水蒸气凝结后，就形成了云。温度、湿度、风速和稳定性条件决定了云彩的外观和大小。

The weather in Copenhagen is overcast with gray clouds two-thirds of the year. The urban space, CLOUD, reflects this situation by using the gray shades of color in its paving and through its big and open surface to allow for the water vapor, fog, and mist that often characterize the harbor in the spring and summer months.

哥本哈根全年有三分之二的时间是笼罩在阴雨天下的。城市区域，CLOUD，通过运用灰色阴影来反映这种情况；在春夏季节，往往通过宽大开放式的表面让水气、雾气和薄雾来塑造装饰海港。

All in all, CLOUD provides Copenhagen with a sensuous and intimate urban space that not only provides local acclimatization and amenity values, but also allows the residents of Copenhagen to enjoy some of their city's most characteristic traits; clouds, rain and mist.

总而言之，COULD 设计为哥本哈根提供了感性而亲切的城市空间，这里不仅能够提供适合当地的健康舒适的环境，而且能够让哥本哈根的市民享受到自己城市独具特色的一面：云、雨和雾。

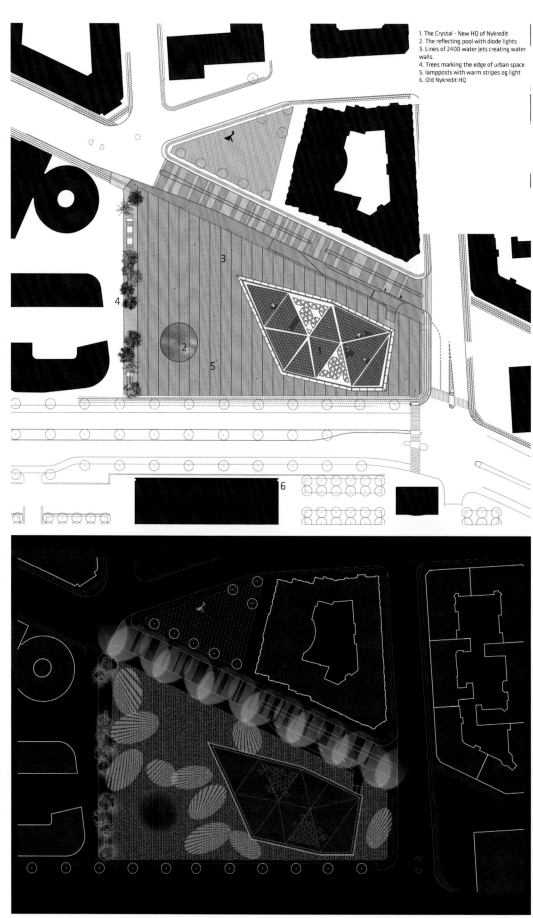

1. The Crystal - New HQ of Nykredit
2. The reflecting pool with diode lights
3. Lines of 2400 water jets creating water walls.
4. Trees marking the edge of urban space
5. lampposts with warm stripes og light
6. Old Nykredit HQ

Fredericia C — Temporary Park

Fredericia C — 临时公园

LOCATION: Fredericia, Denmark
项目地点: 丹麦 腓特烈西亚

AREA: 140,000 m²
面积: 140 000 平方米

COMPLETION DATE: 2011
完成时间: 2011 年

ARCHITECT: SLA
建筑设计: SLA

DESIGN COMPANY: SLA
设计公司: SLA

Fredericia C — Temporary Park

Throughout the 20th century, Fredericia's harbour was dominated by heavy industry. This is now starting to change. In 2004, one of the harbour's most prominent industries, Kemira Grow-How, ended operations and demolition began. Realdania (a foundation working to improve the built environment) bought the Kemira site in June 2008, together with other sites along the harbour. Realdania and the City of Fredericia have since formed Fredericia C P/S in a joint venture to regenerate the harbour area. SLA was hired to help this long regeneration process. Thus the concept of the "temporary park" was born.

在整个 20 世纪，腓特烈西亚港口都以重工业为主导产业。现在，这里开始发生了变化。2004 年，Kemira Grow—How，港口最突出的产业之一，终止了运营业务，开始进行拆迁。Realdania（从事改善建筑环境基础工作的公司）于 2008 年 6 月购买了 Kemira 及其他一些沿海港的区域。Realdania 和腓特烈西亚城联合成立了一家合资企业——Fredericia C P/S，用以重建和净化海港区域。所以，他们聘请 SLA 协作此项长期的重建过程。因此，"临时公园"的概念由此而生。

Since the summer of 2009, a part of the harbor had been opened to the public and had been used for everything from picnics and fishing to events such as concerts, theatre, and vintage car shows. By spring of 2010, the Kemira site had been completely cleared, and construction work for the temporary park project went underway. The project was completed in phases, scheduled to open in July, August and September 2010.

自 2009 年的夏天以来，港口的一部分已经向公众开放，并且已用于从野餐到钓鱼等一系列活动，并可承办戏剧演出、音乐会、老式汽车展等。直到 2010 年春，Kemira 区域已经得到了全面的净化治理，此外，临时公园项目的建设工作也正在进行中。项目分阶段完成，于 2010 年七、八、九月开放。

The idea behind the project is to establish a robust framework, which can subsequently be allocated with different functions. The design of the framework is based on historical maps of the area. The physical expressions of the frames are determined by planting, materials, lighting, and the juxtaposition of these elements. The expression is raw and simple, as it should be easy to build and remove the temporary arrangement as the new urban area takes shape.

该项目背后的构想是要建立一个强健的体系框架，随后可以配置各种不同的功能进去。该框架的设计是基于该地区的历史地图的。构架的有形表现形式是由花圃、材料、照明以及这些元素的并置搭配所决定的。这些表现形式是简朴而无雕琢的，而且正因为基于刚成形的新城区，这里应该方便建设和移除临时的布局安排。

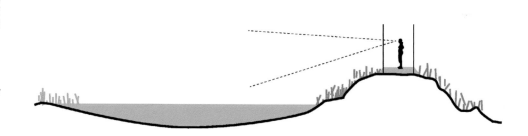

The framework consists of various temporary pavings and spatial elements which create great opportunities for a range of different activities regarding "health and exercise". The included elements are based on a great involvement of the citizens of Fredericia. Spaces have also been left open within the framework so that these can be filled in over time as new needs and wishes arise.

该体系框架是由各种各样的临时路径和空间元素组成的。这些路径和空间元素能够为一系列与"健康和运动"相关的活动创造非常多的机会。这里有多种不同的与"健康和运动"有关的主题活动项目。其中所运用的想法和元素都是基于腓特烈西亚市民大力支持和积极参与之上的。因为建设体系框架内的空间也都是处于开放状态的，所以随着时间推移而产生的新需求和新愿望也将会添入到整个建设中去。

North West Park, Copenhagen

哥本哈根西北公园

LOCATION: Copenhagen, Denmark
项目地点：丹麦 哥本哈根

AREA: 35,000 m²
面积：35 000 平方米

COMPLETION DATE: 2010
完成时间：2010 年

TEAM: Stig L. Andersson, Rasmus Astrup, Friedemann Rüter, Thomas Kock, Salka Kudsk, Signe Høyer Frederiksen, Michelle Nielsen—Dharmaratne, Cecilie Milsted Lind, Simone Maxl and Martin Søberg
团队：Stig L. Andersson, Rasmus Astrup, Friedemann Rüter, Thomas Kock, Salka Kudsk, Signe Høyer Frederiksen, Michelle Nielsen—Dharmaratne, Cecilie Milsted Lind, Simone Maxl 和 Martin Søberg

COLLABORATORS: Author Jan Sonnergaard, Lemming & Eriksson, 2+1
合作者： Author Jan Sonnergaard, Lemming & Eriksson, 2+1

DESIGN COMPANY: SLA
设计公司：SLA

North West Park, Copenhagen

North West is the most culturally diverse quarter in Copenhagen with almost twice the amount of non—western immigrants as in the rest of the city. As a result, the quarter has had its share of social and economic troubles: the inhabitants of North West are poorer, more often unemployed, and live in smaller public—housing apartments than their fellow Copenhageners. North West itself is a grey area with little public life and is known for its high rates of crime and pollution. It is not a coincidence that North West is popularly known to the rest of Copenhagen as "North Worst".

西北区是哥本哈根文化最多元的一个城区，所拥有的非西方移民的数量几乎是哥本哈根其余地区的两倍。因此，该区域是存在一定的社会和经济问题的：西北区的居民越发贫穷，而且经常失业，相比其他哥本哈根人，西北区的居民居住在较小的公共房屋里。西北区本身就是一个拥有很少公共生活的灰色地带，这里是因犯罪率和污染率高而闻名的。由此可见，西北区被俗称为"差北区"并不是个巧合。

With SLA's design of their new public park, the people of North West were handed a unique opportunity to reinvent themselves. The goal was to transform a vast, fenced, empty space (formerly the contaminated area of the city's bus terminal and garage) into a space where people of all cultures, nationalities and ages can meet. With a magic mix of lights, colors, trees, poetry and even small mountains, new adventures and stories were introduced to North West; a place where the quarter's residents with all their differences and colors would find space to unfold themselves and flourish all year round.

SLA 的新式公共公园设计，为西北区的人们提供了一个重塑自身的难得机会。设计的目标是将一片广阔的、闭塞的、贫乏的空间（原城市巴士站和车库的污染区）转变成一片适宜所有不同文化、民族和年龄的人群生活的温馨天地。其中融入了灯光、颜色、树木、诗歌、甚至还有小山，再加上新颖的冒险经历和故事的神奇组合，那么，随着这一系列的神奇组合引入西北区，当地所有不同文化和肤色的居民都将会找到适合展现自己的空间。在这里，全年都充满着欣欣向荣、蓬勃发展的氛围。

Under the title "1001 trees", the park consists of four simple, but effective elements; trees, paths, light, and cone—shape mounts. These elements create order and coherence between the park's many different parts. All four elements are distinct features of the park. Their simple, but varied compilation creates a sequel of changing spaces and corners with altering atmospheres and feelings, which differentiate the park from the area's grey and fragmented environment.

在"1001 树林"的主题下，公园包括了四个简约、有效的元素：树木、道路、光线和锥形座椅。这些元素为公园很多不同的部分之间创建了某种秩序和连续性。这四要素都是公园鲜明的特色。他们虽然简约，但是通过不同的整编搭配，创造了一系列随着气氛和感觉更迭而变化的空间和角落。这些元素的运用和设计让公园与城市零散、灰色的环境迥然不同起来。

The trees of the park were selected to have a variety of different geographical origins, creating a beautiful and exciting encounter between traditional species of the Danish latitudes and exotic species from around the world. This reflects the multicultural nature of the North West area and is also intended as a critical commentary on the Danish restrictive immigration laws.

公园林木要根据不同的地理起源来进行选择，在丹麦纬度位置的传统种类和世界各地的外来物种之间，创造出一种绚丽并令人兴奋的融合。这里反映了西北地区的多元文化性质，也成为丹麦限制移民法的一个重要评注。

North West as an area is going through changes. With SLA's design, the park is a symbol and leader in this positive change. The North West Park is providing a run—down neighbourhood with an open park that protects and mirrors the diversity and changeability of the area. The park reflects the diversity of the area, the adventurous spirit, the strong engagement, the need to get lost in the crowd and to seek into the quiet realm of reflection. The park embraces the need for everyone to feel welcome, while in quality competes with the finest urban spaces of Europe.

西北区是一个正在历经变迁的区域。基于 SLA 的设计，公园是这次积极变革的一个标志和导向。西北公园为破败的社区提供了一个开放式的公园，用以保护和反映该地区的多样性和变化性。公园会反映出该地区的多样性、冒险精神、强烈的参与性、融入人群的需要以及寻求沉思的宁静境界。公园会满足每一位游人的需求，让人们有宾至如归的感受，同时在服务质量上，公园将会与欧洲最好的城市公园相比肩。

The City Dune / SEB Bank—Acclimatized Urban Space in Copenhagen

城市沙丘／瑞士 SEB 银行 — 适应哥本哈根的城市空间设计

LOCATION：Copenhagen，Denmark
项目地点：丹麦 哥本哈根

AREA：7，300 m²
面积：7 300 平方米

COMPLETION DATE：2010
完成时间：2010 年

ARCHITECT：SLA
建筑师：SLA

DESIGN COMPANY：SLA
设计公司：SLA

The City Dune / SEB Bank — Acclimatized Urban Space in Copenhagen

The City Dune, as the urban space quickly came to be called, is made of white concrete, borrowing its big, folding movement from the sand dunes of Northern Denmark and the snow dunes of the Scandinavian winter. The folding movement and the contour of the terrain not only handle functional and technical demands from drainage, accessibility and lighting to plantation and the creation of a root-friendly bearing layer. It also offers a variety of routes for customers and employees of SEB as well as ordinary Copenhageners, creating an ever-changing urban space.

这个城市空间是由白色混凝土浇注而成的，很快地就被人们称为城市沙丘，灵感来源于丹麦北部的沙丘和北欧斯堪的纳维亚冬季的雪丘。层层叠叠的混凝土和地形的轮廓线设计不仅是为了排水功能、辅助功能、人工林的可达性和照明功能的需要，它还为 SEB 的顾客、员工以及哥本哈根的市民提供了多条线路，创建了一片不断变化中的城市空间。

To fully experience The City Dune, one has to physically move through it. When passing through the area, the space evolves and opens up in different directions, creating new spatial connections in the process. When ascending from Bernstorffsgade, the space gradually unfolds as you walk along the 300 m long and winding incline. Looking back against the city, the buildings frame a solid cut of Copenhagen. The ascent from Kalvebod Brygge is shorter and steeper. Here one will soon appreciate a splendid view of the harbor.

如果要充分领略城市沙丘，人们一定要亲身穿越它。经过该地区时，空间会向不同方向铺开，在展开过程中，创建新的空间连接。空间从 Bernstorffsgade 开始上升，当你沿着曲折的斜坡行走 300 米，空间就会逐渐地展开来。回顾这座城市，建筑物成为了哥本哈根一个形象立体的标签。Kalvebod Brygge 的上升变得更狭窄而且更陡峭。在这里，人们可以很快地领略到海港的壮丽景色。

Acclimatization is the most important principle in the design of The City Dune. Through the folding movements of the concrete, the surface reflects as much of the incoming sun's radiation as possible, thereby creating a cooler microclimate during the hot months of the year. This is further enhanced by 110 water atomizers emitting out moist air, spread by the wind. The result is the experience of being in the middle of the lush Scandinavian nature. Narrow drains lead the rainwater from the concrete surfaces into two large rainwater tanks. From here it is pumped to the plantation and the water atomizers through a fine—meshed network of tubes. As such, no rainwater ends up in the sewers or on the roads.

在这个城市里程碑式的项目中，环境适宜性是最重要的原则。层层叠叠的混凝土最大化地反射了表面的太阳辐射，为一年中最热的几个月创造相对凉快的小气候。通过110个水式喷雾器喷射出潮湿的空气，并由风力将潮湿的空气传播蔓延开。其效果是使人们有置身在郁郁葱葱的斯堪的纳维亚大自然中的感觉。可见，场地上的人工林和110个喷雾器加强了凉爽的感觉。狭窄的排水沟将混凝土表面的雨水引导到两个大型的集水池中。这些雨水经过一个精密的管道系统，被用于灌溉树木和引入喷雾器中。因此，没有任何雨水最终进入下水道或马路而被浪费。

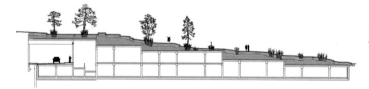

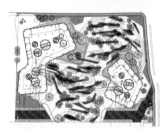

The trees and herbaceous borders are placed in fissures between the horizontal planes. Both deciduous and evergreen plantation has been utilized to achieve the metabolism of water throughout the year in addition to enhancing the microclimatic environment with wind and shelter. The trees and plantations are not arranged to emulate nature. It is a new manner of seeing and experiencing nature in the city. The ambition is to create an urban view of nature through a design that clarifies the presence of nature as a process, while simultaneously supporting acclimatization and other functional conditions.

局部下陷的水平面裂缝界定了树木和草坪的边界线。为了达到全年水分的新陈代谢以及加强小气候，除了通过风力和覆盖物来实现外，场地上还种植了落叶树和常绿树。林地和种植区的设计并不是为了模拟自然，而是在城市里给人一种新的看待和体验自然的方式。这样做的目的在于创造一种城市自然景观，呈现自然过程，与此同时，提供适宜环境及其他的功能。

All in all, the City Dune not only provides acclimatization and utility through the sustainable use of concrete and plantation; it also gives a much needed recreational value to a part of Copenhagen long neglected by city planners.

总而言之，这个城市里程碑不仅是哥本哈根第一个100%可持续地利用混凝土和人工林来营造城市空间的项目，它也提供了被城市规划师长期忽略的迫切需要的游憩功能。

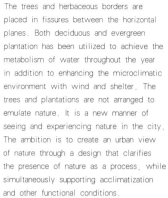

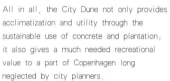

The Office of James Burnett

OUR HISTORY

The Office of James Burnett was founded in 1989 in Houston, Texas with the primary focus of creating landscapes that transform perspective and evoke emotional responses, creating unique and unforgettable sensory experiences in the garden. Our work is a testament to compositions that reflect a seamless, unified relationship between landscape and architecture. A second office was opened in San Diego, California in 2003.

OUR PROCESS

The Office of James Burnett follows a proven design and planning process in which the client is an integral part. We stay connected, listen and respond to their needs. Through every phase of the design process we stay focused on the site users and their needs. From concept design to construction administration, we know how to work with owners, project managers, architects, contractors and other consultants.

OUR PASSION

Landscape design is our passion — it is who we are, not just what we do — and that passion is reflected in the spaces we create. We consider all facets of your project, beginning with the first impression and concluding with the finest detail.

Brockman Hall for Physics at Rice University

莱斯大学物理学布罗克曼大厅

LOCATION: Houston, USA
项目地点: 美国 休斯顿

AREA: 10,312 m²
面积: 10 312 平方米

PHOTOGRAPHER: Hester + Hardaway
摄影: Hester + Hardaway

DESIGN COMPANY: The Office of James Burnett
设计公司: The Office of James Burnett

Brockman Hall for Physics at Rice University

Brockman Hall for Physics at Rice University

莱斯大学物理学布罗克曼大厅

LOCATION: Houston, USA
项目地点: 美国 休斯顿

DESIGN COMPANY: The Office of James Burnett
设计公司: The Office of James Burnett

The Brockman Hall for Physics is a 10,312 m² facility housing classrooms, laboratory space, lecture halls and administrative offices for the Physics Department as well as physicists from the Electrical and Computer Engineering Department at Rice University in Houston, Texas. Driven by Rice University's belief that some of the most important moments on campus are moments of informal discussion and debate outside classroom, the design of the building and landscape seeks to provide a multitude of spaces for lively and inspiring conversation. Sheltered from the sun by the building overhead, a ground-floor courtyard features a reflecting pool, raised ipe terrace and an enhanced plaza with movable furniture. As the design developed, the Office of James Burnett was also asked to redesign the "Courtyard of Science", an interstitial space between the wings of Brown Hall to the south. A grove of Honey Mesquites organizes the space and intimate decomposed granite courtyards with movable furniture create a number of social spaces.

物理学布罗克曼大厅是一座面积为 10 312 平方米的建筑，是为位于德克萨斯州休斯敦市莱斯大学的物理系及电气与计算机工程系的物理学家们而建的，里面容纳了教室、实验场所、演讲大厅和行政办公室。课堂外非正式的讨论和辩论是校园活动中最重要的部分之一，这是莱斯大学的理念，在这一理念的驱动下，整个建筑及景观的设计都致力于为热烈而精彩的对话提供大量的空间。一楼的庭院被上方的楼遮挡着，里面有倒影池、突起的重蚁木平台和配有可移动家具的高级广场。随着设计的展开，the Office of James Burnett 设计公司还被要求重新设计"科学园"。它位于南侧布朗大厅的两翼之间，一片牧豆树把整个空间排布得井然有序，里面的小庭院铺有风化的花岗岩，放置了可挪动的家具，十分温馨，营造出了许多的社交空间。

The Brockman Hall for Physics is a new 10,312 m² facility at Rice University. Gathering faculty and researchers that were formerly located in several buildings across campus, Brockman Hall is the new home for physics research at Rice. The building and landscape aid this research both by providing a home for the laboratories, classrooms, and offices and by supplying informal gathering spaces to foster conversation, debate, and cross pollination of ideas.

物理学布罗克曼大厅是莱斯大学一座新建的面积为 10 312 平方米的建筑，它将以前分散在校园内的物理学教职工和研究人员聚集到了一起，成为莱斯大学物理研究的新家园。它的建筑和景观为实验室、教室和办公室提供了场所，辅助了该学科的研究，为非正式的聚会提供了场所，促进了对话、辩论和交流。

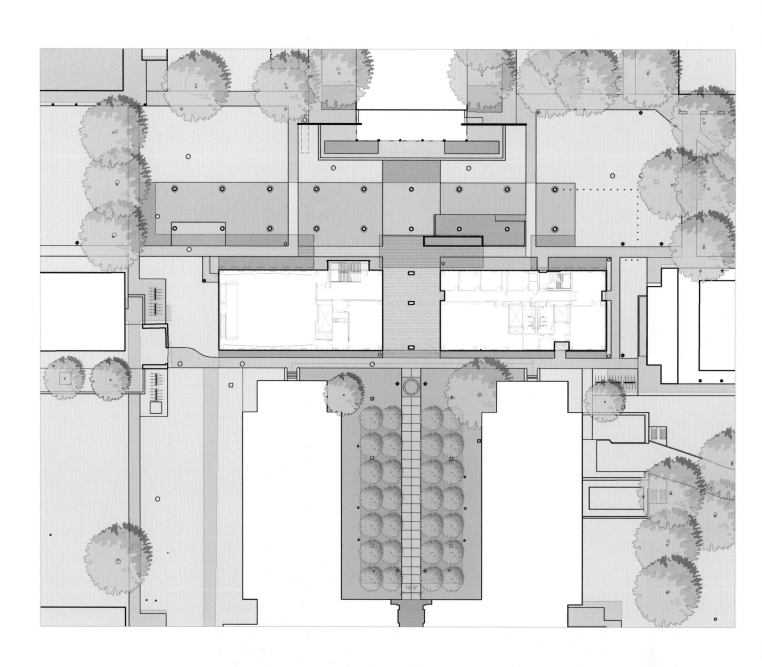

Sunnylands Center and Gardens

Sunnylands 中心花园

LOCATION: CA, USA
项目地点：美国 加利福尼亚

AREA: 809,371 m²
面积：809 371 平方米

PHOTOGRAPHER: Hester + Hardaway, The Office of James Burnett, Dillon Diers
摄影：Hester + Hardaway, The Office of James Burnett, Dillon Diers

DESIGN COMPANY: The Office of James Burnett
设计公司：The Office of James Burnett

Sunnylands Center and Gardens

An extension of the 809,371 m² desert retreat of a publisher, diplomat and philanthropist Walter Annenberg, the Sunnylands Center and Gardens is an interpretive center that celebrates the architectural and cultural legacy of the historic estate. Inspired by the Annenberg's extensive collection of impressionist artwork, the landscape architect painted a living landscape that respects the character of the Sonoran Desert and demonstrates a new ecological aesthetic for landscapes in the arid southwest.

Sunnylands 中心花园是个占地 809 371 平方米的沙漠度假地的扩建项目，为著名出版人、外交家和慈善家 Walter Annenberg 所有，它宣扬和阐释了这座历史园林所秉承的建筑和文化遗产。景观设计师从 Annenberg 广泛收集的印象派艺术品中得到灵感，描绘出了一幅符合索诺兰沙漠特征的生活景观，在美国贫瘠的西南部呈现一种新的生态美学。

Sunnylands Center and Gardens in Rancho Mirage, California is an extension of the 809,371 m² desert retreat of the publisher, diplomat and philanthropist Ambassador Walter Annenberg and his wife Leonore. The Annenbergs commissioned the California modernist architect A. Quincy Jones to design their estate in the desert in 1963. In 2006, the Annenberg Foundation commissioned the design team to develop an interpretive center that tells the story of the Annenberg's contributions to the cultural, artistic and architectural history of America.

Sunnylands 中心花园位于加利福尼亚州 Rancho Mirage 市，是一个沙漠度假地的扩建项目，该地产占地 809 371 平方米，为著名出版人、外交家及慈善家 Walter Annenberg 大使和他的夫人 Leonore 所有。在 1963 年，Annenberg 夫妇委托加利福尼亚的现代主义建筑设计师 A. Quincy Jones 在沙漠中为其设计了这个度假之地。在 2006 年，Annenberg 基金会授权该设计团队开发一个中心，以展示 Annenberg 先生对美国文化、艺术及建筑历史的贡献。

Because of its location in the desert, sustainability figured prominently into discussions about the nature of the project. Originally conceiving an extension of the golf-course landscape of the estate, the design team and the client came to agree that it was in everyone's best interests to implement the most advanced efforts in sustainability.

由于该中心地处沙漠里，所以在探讨项目性质时，可持续性成为重点的考虑。最初的构想是对该度假地的高尔夫球场景观区进行扩建，设计团队和客户都认为在可持续性上进行高水平的投入和付出是最重要的。

In addition to the selection of regionally appropriate plants, the project features restored desert habitat, a high-efficient capillary irrigation system, a soil moisture monitoring system, a on-site storm water retention system, geothermal wells, a significant photovoltaic array and an on-site green waste recycling program. The Center is a pilot project for the Sustainable Sites Initiative, is pursuing LEED Gold Certification and uses approximately 20% of its water allocation from the Coachella Valley Water District. The project also proactively meets the specifications and requirements for the use of reclaimed water five years ahead of the implementation of Rancho Mirage's citywide initiative.

除了选择适应的区域性植株之外，该项目的特色还包括充满生机的沙漠栖息地、高效的毛细灌溉系统、土壤湿度监测系统、园内雨水保留系统、地热水井、太阳能电板，以及园内绿色环保的垃圾回收点。该中心是可持续建筑倡议活动的试点工程，只使用 Coachella 山谷供水区所供应水量的 20%，目前正努力争取获得节能绿色建筑金级认证。该项目也积极主动地达到再生水的规格和要求，比 Rancho Mirage 市倡议的实施计划提前了五年。

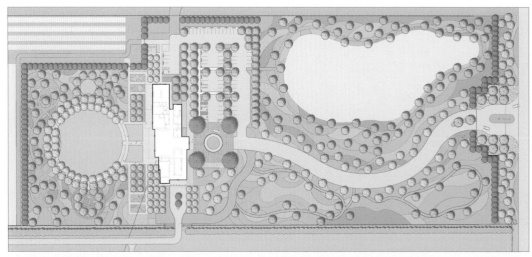

Working closely with Mrs. Annenberg and the Annenberg Trust, the landscape architect developed a scheme that an orderly, geometric composition closest to the building becomes progressively more organic towards the edges of the grounds. Inspired by the owner's painting, "A Wheatfield with Cypresses" by van Gogh, the landscape architect sculpted the earth and used plants in a painterly fashion across the landscape. Trees were carefully positioned throughout the site to ensure that ample shade was provided and great care was given to the visual composition of understory plantings.

景观设计师与 Annenberg 夫人和 Annenberg 信托公司密切合作，制定出一个方案：最靠近楼的地方是规则的几何布局，逐渐地自然地向边缘过渡。该地产所有者收藏的一幅梵高画作——柏树旁的麦田，给了景观设计师灵感，对土地进行雕琢，把整个景观里的植物布置得像画家画的一样。园区里的树木种植都经过精心的安排以确保人们得到充足的阴凉，同时对矮层植物所构成的视觉画面也给以足够的重视。

The Campus at Playa Vista

Playa Vista 校园

LOCATION：CA，USA
项目地点：美国 加利福尼亚

AREA：258,999 m²
面积：258 999 平方米

PHOTOGRAPHER：Hester + Hardaway，The Office of James Burnett，Dillon Diers，Kerun Ip
摄影：Hester + Hardaway，The Office of James Burnett，Dillon Diers，Kerun Ip

DESIGN COMPANY：The Office of James Burnett
设计公司：The Office of James Burnett

The Campus at Playa Vista

The Campus at Playa Vista

Located on the former site of the Howard Hughes Aircraft facility, the Campus at Playa Vista is a 258,999 m² office campus that engages the historic industrial fabric of the property. Preservation of the main aircraft hangar, which once housed the famed Spruce Goose, as well as the other ancillary buildings established the LEED accredited architectural and landscape architectural vernacular for the development.

Playa Vista 校园坐落于霍华德休斯飞机工厂的旧址，是一块占地 258 999 平方米的学校办公区，利用了场址上具有历史意义的工业建筑格局。曾经存放著名的 Spruce Goose 飞机的主库房还有其他的辅助建筑被保存了下来，使该地的开发在建筑和景观上具有地域特色，并获得了绿色建筑协会的认证。

A central 36,422 m² park consisting of sport courts, playground, soccer field, botanical gardens, water features, and a bandshell, serves as the social hub for the campus. Each parcel has park-front access or direct views to the central green providing a strong relationship between architecture and landscape. Richly landscaped courts and roof gardens are integrated with the proposed and historic buildings providing tenants with easy access to the famed outdoor environment of Southern California.

一个 36 422 平方米的中心公园由活动场、操场、足球场、几处花园、几个水景，和一个户外音乐台组成，充当校园的社交中心。每部分空间在公园前侧都有入口，都能直接看到中心的草坪，草坪在建筑和景观之间建立起紧密的联系。景致丰富的庭院和屋顶花园与拟建的历史性的建筑融为一体，为房客提供便利的途径去体会加利福尼亚南部著名的户外环境。

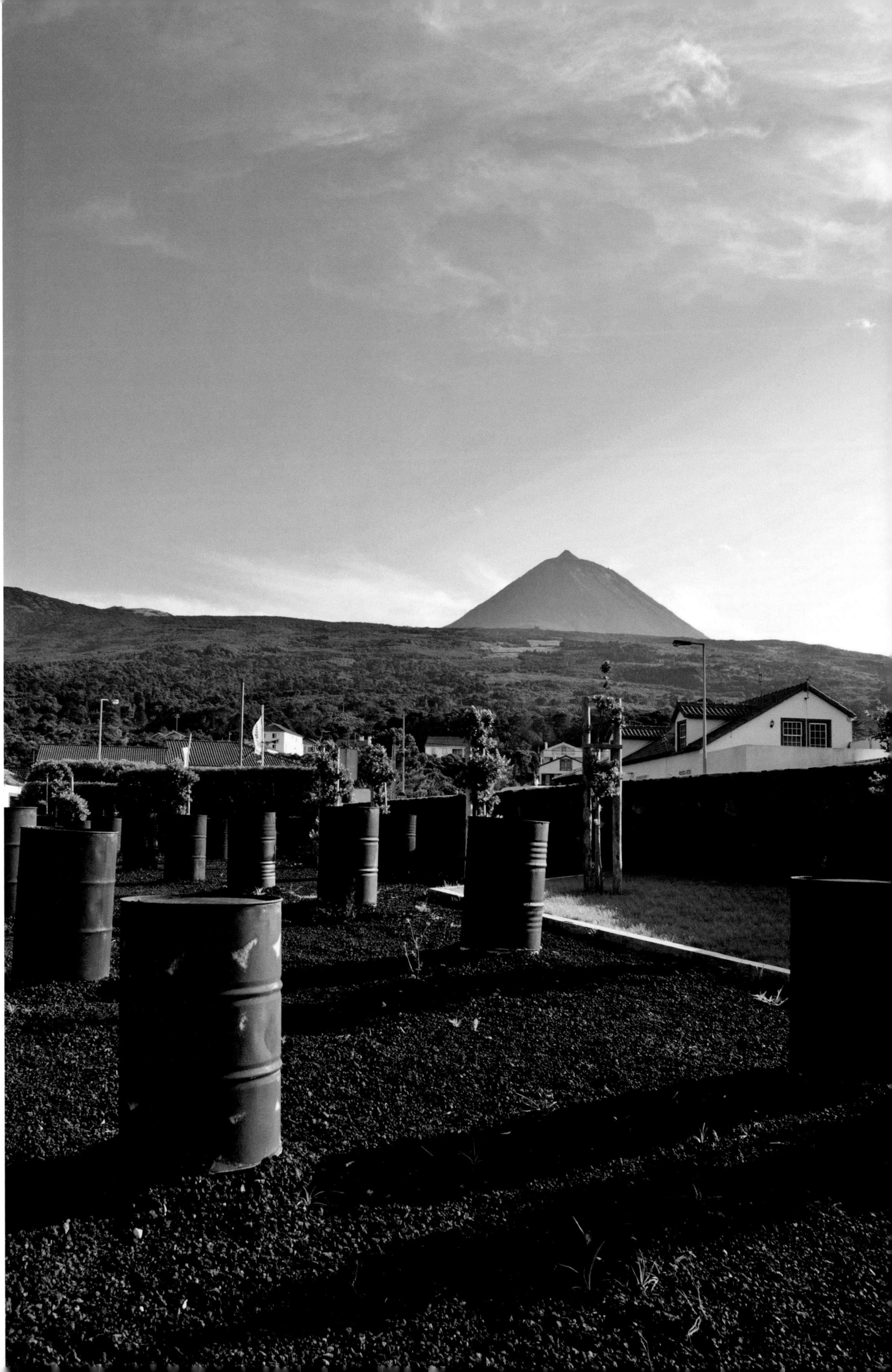

ARPAS — Arquitectos Paisagistas Associados, Lda

A ARPAS - Arquitectos Paisagistas Associados, Lda. é uma empresa de arquitectura paisagista com extensa experiência nas suas múltiplas vertentes, desenvolvendo projecto ou planeamento em espaço urbano, rural ou costeiro.

Foi fundada em 1992 pelo Arq. Pais. Luis Cabral decorrente da sua experiência como projectista desde 1981.
A equipa integra ainda os arquitectos paisagistas Vasco Costa Simões, Gonçalo Saúde Pereira, José Luis Coelho e Maria Maltez.

A criação e recuperação de paisagens centra-se, para nós, no engrandecimento da identidade do sítio, na sustentabilidade do espaço, na simplicidade do gesto, na adequação ao programa e ao território.

Privilegiamos fruição e inovação.

Whaling Industry Museum

捕鲸博物馆

LOCATION：Portugal
项目地点：葡萄牙

COMPLETION DATE：2010
完成时间：2010 年

DESIGN COMPANY：ARPAS — Arquitectos Paisagistas Associados, Lda
设计公司：ARPAS — Arquitectos Paisagistas Associados, Lda

Whaling Industry Museum

Whaling Industry Museum

捕鲸博物馆

LOCATION：Portugal
项目地点：葡萄牙

COMPLETION DATE：2010
完成时间：2010 年

DESIGN COMPANY：ARPAS — Arquitectos Paisagistas Associados, Lda
设计公司：ARPAS — Arquitectos Paisagistas Associados, Lda

The Whaling Industry Museum occupies an abandoned factory of whale derivatives from 1946 to 1984. The present renovation was undertaken in 2011.

捕鲸博物馆的前身是一个鲸产品加工厂，这个1946年建造的工厂在1984年的时候遭到废弃。2011年人们将这个废弃的工厂重新装修，改建成捕鲸博物馆。

The Museum is closely related to the port, dipping into the dock through ramps located a few meters away used in the past to unload the boats and dismantle the cetaceans.

博物馆离港口很近，沿着坡道伸向几米之外的码头。过去人们在这里卸船和解剖处理捕获的鲸鱼。

The courtyard of the factory was once a functional storage area, wide and open, with plenty of barrels of oil and instruments.

加工厂里有个宽敞、开阔的院子，以前是用来储物的，里面放了很多油桶和工具。

This renovated courtyard is now a multipurpose area, not just to frame for the building facade, but also to be used as a living space, allowing events in continuity with the exhibition.

装修后的庭院变成了一个多功能区。它不仅为博物馆墙面提供框架支撑，还可以用作活动区，这样如果需要配合馆内展览举办相应活动的话，就有活动场所了。

Framing this enclosure paved with red volcanic local stone, in transition to a surrounding meadow, we installed wooden benches along the tree line and a platform that can serve as a stage or as seating area.

庭院周边铺的是当地产的一种红色的火山石，外围是草地。在火山石和草地之间是沿着树木安放的一排木头长凳，还有一个兼作舞台和休息区的平台。

Behind this stage, we locate a "whale industry memorial". This memorial is an installation consisting of six magnificent large reservoirs and autoclaves surrounded by former rusty oil barrels disposed in a serial rhythmic formal design, forming pathways with light and shadow. This memorial is enhanced by the black volcanic stone of the pavement.

舞台后面是"捕鲸纪念馆"。纪念馆内有6个巨大的蓄水池和高压蒸气灭菌器。周围放置了一些旧油桶。由于年份比较长了，这些油桶已经锈迹斑斑，但它们的摆放却很有韵律感，在通道上投影一种奇妙的效果。纪念馆的装饰采用了黑色火山石，更具厚重感。

The framing west wall was raised 40 cm with local existing stone to favour the sense of intimacy without blocking the view of all the other facades. The end wall is 1.2 m in height.

西墙加高了40厘米，当地石头的采用让博物馆显得更为亲切，而且这个增高设计并未挡住其他方向的景致。端墙的高度为1.2米。

At the periphery of the enclosure, near the walls, we proposed a grove of Metrosideros, with some Casuarinas diluting the surrounding buildings with less architectural quality, and providing shelter and shade for visitors. Openings will be left in that grove to allow visualisation of the whole museum facade in its full scale and size, now free from the proximity of the abandoned barrels.

在庭院四周，靠墙的位置，我们提议种上一丛丛的新西兰圣诞花和常绿灌木来淡化墙体的感觉，并为参观者提供一些阴凉的休息处。树丛中留出了一些开口，从开口处可以看到博物馆的外观全景。从这些角度看过去都见不到那些废弃的油桶。

The central zone is in red lava sand (bagacina) in contrast with the black stone on the buildings facade. This inexpensive local material was chosen for the treatment of such a large area, also for its multifunctional character allowing various types of events and festivities benefiting the museum as a whole.

中间区域铺的是巴嘎西纳山上的红熔岩土，和博物馆外墙的黑色石头形成鲜明的对比。中间区域面积较大，选用当地的这种材料相对便宜一些。而且这种材质的地面比较结实，可以适应博物馆举办的各种活动和节日庆典。

The ramps, stone benches and trees constitute the renewed front of the entrance favouring a pedestrian approach, creating a distance to parking space.

博物馆入口处的坡道、石凳和树在为行人提供便利的同时，也有效地将停车区分隔开了。

The entrance was modified to provide wider views of the facade, the ramps and their industrial gears and mechanisms. The new pavement in concrete paving slabs enhanced the machinery that pulled the whales to the shore.

博物馆的入口处经过了改良，由此可以更好地看到外墙、坡道和捕鲸设备。地面铺设的厚石板也有助于拖鲸机器拖鲸上岸。

John Feldman

CEO, Founding Principal

A native of Los Angeles, John Feldman studied at the College of Architecture and Environmental Design at California Polytechnic State University, San Luis Obispo, where he received his Bachelor of Science degree in Landscape Architecture. Before beginning his professional career he embarked on an extensive overseas study program traveling throughout China and Southeast Asia. Through both his own individual study and his collaboration with professional offices and universities, he sought to explore, in depth, the social and cultural impacts on architecture, urban planning and the natural environment.

Feldman has been involved in a wide range of project types, including commercial retail, street improvements, museum and institution, public open spaces, multi—family housing/mixed—use planning, residential gardens and estate master planning. His skills reflect the diversity of the projects and his ability provides expertise in design, public relations, technical problem solving, scheduling and budgeting issues.

Feldman enjoyed his tenure at some of the most prolific design firms in Los Angeles. As Director of the Landscape Studio at KAA Design Group, he directed all aspects of design, management, and construction administration for the range of opportunities at the firm. Feldman credits having honed his skills in design and theories in "regional contextualism" at the firm Nancy Goslee Power and Associates. In addition to strong business skills developed while having formerly operated the firm Garness/Feldman—Architecture + Landscape, Feldman brings extensive international design experience from his completed landscape projects in many countries around the world.

Licensed in the States of California and Hawaii, Feldman enthusiastically leads Ecocentrix, Inc. with vigor, vision, and evolved paradigms, with resulting design investigations spanning tradition to progressiveness—whereever the firm's work takes them.

Ecocentrix

Ecocentrix was founded on the fundamental premise that the quality of the experience and function of landscapes is achieved by understanding inherently "what is"and "what is wanted", and that quality of life is a reflection of the quality of the landscape. It has remained in our charge to maintain a keen understanding of our client's individual and collective interests and lead each project with an evolved vision for how these may manifest within.

The firm's work is rooted in investigations of residential estate and resort style living. Our clients are characterized by their culturally rich backgrounds and sophisticated design tastes, ranging from traditional to contemporary, and whose personal lifestyles and histories include a diverse range of travel and worldly explorations. This has availed us the opportunity for continued studies in the area of luxury living, and to enjoy work internationally.

Whether residential resort or resort hotel, our approach to site planning and amenity design bears a similar thread. We artfully interact with nature by thoughtfully manipulating natural and constructed form, recognizing that the art of landscape is in the interaction of human and nonhuman nature. It is in this way that we may accentuate and amplify space.

Our body of work exemplifies great stylistic range and restraint produced with consistently high quality. Our projects are immediately mood altering, celebrating the sensual and tactile temperament that is the fabric of landscape.

Our design creates the ground for celebrating the cycles of all life, and is the foundation of regional identities enveloping cultural distinctions. It reinforces what is powerful and enhances what is weak. Ecocentrix endeavors to "Enrich Life Through Design".

Beverly Hills Retreat

贝弗利山隐居所

LOCATION: CA, USA
项目地点：美国 加利福尼亚

AREA: 1,335 m²
面积：1 335 平方米

COMPLETION DATE: 2009
完成时间：2009 年

PHOTOGRAPHER: Manolo Langis
摄影：Manolo Langis

DESIGNER: Ecocentrix Landscape Architecture
设计师：Ecocentrix Landscape Architecture

DESIGN COMPANY: Ecocentrix
设计公司：Ecocentrix

Beverly Hills Retreat

Beverly Hills Retreat
完成时间：2009 年

PHOTOGRAPHER: Manolo Langis
摄影：Manolo Langis

This small single-story contemporary residence is nestled in the hills of Beverly Hills, close to the well-known Mulholland Drive. It was originally constructed 50 years ago for a young actress who was a mistress of the famous Howard Hughes.

这个单层的现代小居所置身于贝弗利山中，挨着著名的马尔霍兰大道。它最初是在50年前为一个年轻的女演员建的，她是那个著名的霍华德·休斯的情人。

Its current resident purchased the home 40 years ago who then commissioned the first landscape architect's design at that time. Designed by a well-reputed landscape architect of the period, the design had become dated and the materials had decayed badly. It was our charge to evolve the landscape into a contemporary and simple vernacular that would better match the home's wonderful interiors while also functioning better.

现在的居住者在40年前买了它，然后在那时找来建筑师进行了首次景观设计。当时那是有名的景观建筑师设计的，但它已经过时了，建筑材料腐烂严重。我们的任务是把这一地方开发成一个现代、简洁、具有地方特点的景观，使其更好地与华丽的室内装修搭配，并且发挥出更好的功能。

We opened hillside views, reconstructed the pool and surrounding hardscape, lengthened walls, and incorporated highly graphic plants in long swaths to accent long lines of the property. A new outdoor kitchen boasts all modern conveniences and is well used by the client who hosts frequent gatherings.

我们开发了山坡景色，重建了水池和周围硬件景观，把墙延长，在狭长地带加入造型植物来突出住所的长线条。一个新的户外厨房装备了所有的现代化便利设备，很气派，在客户频繁举办的聚会中发挥了良好的作用。

Encino Hills Spa Residence

恩西诺山 spa 居所

LOCATION：CA，USA
项目地点：美国 加利福尼亚

COMPLETION DATE：2011
完成时间：2011 年

PHOTOGRAPHER：John Feldman
摄影：John Feldman

ILLUSTRATIONS：Manolo Langis
3D 插画：Manolo Langis

DESIGNER：Ecocentrix Landscape Architecture
设计师：Ecocentrix Landscape Architecture

DESIGN COMPANY：Ecocentrix
设计公司：Ecocentrix

Encino Hills Spa Residence

This home's architecture is modest in its clean lined design, yet hosts fabulous contemporary interiors. The client desired that the new garden and handscape renovations would complement the interior design and make for an inviting graphic, sculptural, and spa—like experience while they entertain guests in their much enhanced home.

这个住所的建筑设计线条简洁流畅，外观普通，但室内现代华美。客户渴望新的花园和人造景观的整修与室内设计相辅相成，当在这个改造一新的家里招待客人的时候，让人感受到迷人的画面和雕塑感，有享受spa一样的体验。

From the home's open floor plan, when people enter the room, they look through to the rear yard. They are then confronted by a brimming raised spa within the pool that glistens with iridescent glass tile like a "Jewel Box", and also emanates a beautiful trickling sound throughout the tranquil garden. The western sky with colorful sunset is reflected on the generous surface of this luxurious pool. A simple restrained palette of hard and soft materials links this wonderful watery space to the icy blue plant colors in the front entry—dining garden.

从居所开放的平面图上看，当走进室内时，会直接看到后院。然后一个突起的spa出现在眼前，它位于水池里，水流外溢，铺着色彩斑斓的玻璃砖，就像一个珠宝盒一样闪闪发光，而且还发出动听的嘀嗒声，在整个宁静的花园里回荡。西边的天空，伴着多彩的日落，倒映在这个华丽水池的宽阔水面上。软硬材料简单朴素的色彩把这个美妙的水域空间与前面入口餐厅花园里冰蓝色调的植物连接在一起。

Whereas much of the original paving throughout the site was patterned in red brick, its new concrete replacement provides powerfully controlled lines and forms strong architectural planes in this landscape.

整个居所大部分的室外地面最初都是由红砖铺设而成的，带有图案，但现在被混凝土取而代之，线条十分明朗，使这块景观呈现出几个坚固的建筑平面。

Plants are massed together to "paint" their conspicuous locations, and boast sculptural forms that are quite dramatic.

植物一块一块地聚在一起，十分显眼，每一块的外形都被雕琢得独特迷人。

The use of the "Mexican Feather Grass" in its long 100-foot swath beside the pool is intended to catch the delicate breeze and evoke a sense of cooling even on the hottest days. Much like that of native dune grasses that sway by the sea, their movement is quite captivating and evokes relaxation whether viewed from inside the home or laying poolside. The combination of these sensuous grasses and the inviting blue pool is quite iconic in this 1950's retro inspired landscape. A simple and gracious lawn wraps the backside of the pool like a carpeted room, with a singular purple flowering Jacaranda Tree as its grace note. The tree punctuates the lawn on bias to the spa, whose gesture also mimics the pool-enveloped spa.

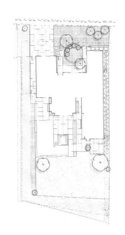

在水池边长 100 英尺的条状地带上种植了墨西哥羽毛草，就是为了扑捉那轻柔的微风，甚至在最热的时候都会带来一种凉爽的感觉。这种草和在当地海边摇摆着的沙丘草很像，摆动起来是那么得迷人，无论从房间里看还是躺在池边都让人感到十分放松。这个 20 世纪 50 年代复古风格的景观充满了灵感，令人愉悦的小草和迷人的蓝色水池结合在一起就是突出的代表。一块简洁而优雅的草坪覆盖在水池的后侧，像房间里的地毯，一棵开着紫色花的蓝花楹树是它的修饰符。这棵树点缀在草坪的对角线上，与 spa 成一条线，它的整个感觉就像是模仿着水池中的 spa。

The entry courtyard dining garden explores a cool color theme with blue succulents, blue grasses, a punch of deep maroon tree foliage, and purple-flowering vines woven on the hung architectural stainless steel lattices. Specimen plants here are used as living sculptures that are back-lighted against a curved wall. This bold garden is theatrical with this being the main stage located immediately adjacent to the formal dining room for all guests to revel in. The tree in this radial layout is kin to the needle on a phonograph and is placed carefully on its outermost ring.

前院入口的餐厅花园展现的主题是冷色调，里面有蓝色的肉质植物、蓝色的草、一片深褐色的树叶，在悬挂的不锈钢格子架上编织着开着紫色花的攀爬植物。这里的标本植物是活的雕塑，以一道弯曲的墙为映衬。这个格调鲜明的花园是那么地惹人注目，和正式的餐厅紧挨着，是所有客人狂欢的主要舞台。在这个放射状的布局中，那颗树好比是留声机上的唱针，被精心地放在最外圈上。

Constructed glass and lumber gates, fixed panels, and garage door are a contemporized interpretation of craftsman style furniture. They intentionally pair with a few of the key interior pieces in the owner's collection and help renew the exterior architecture.

用玻璃和木头制作的门、固定的面板、以及车库的门体诠释了当代的工艺水准，目的就是为了与房屋主人所收集的几件重要室内家具相搭配，有助于展现室外整修的效果。

PV Estate

PV 居所花园

LOCATION：CA，USA
项目地点：美国 加利福尼亚

AREA：8,094 m²
面积：8 094 平方米

PLAN RENDERING：John Feldman
平面图渲染：John Feldman

PHOTOGRAPHER：John Feldman
摄影：John Feldman

DESIGNER：Ecocentrix Landscape Architecture
设计师：Ecocentrix Landscape Architecture

DESIGN COMPANY：Ecocentrix
设计公司：Ecocentrix

PV Estate

LOCATION：CA，USA
项目地点：美国 加利福尼亚

AREA：8,094 m²
面积：8 094 平方米

This 8,094 m² terraced hillside botanical garden is poised high above the Pacific Ocean. It is defined as long views on and off site and is scribed by indigenous on—site excavated stone used for two 100—foot long battered walls. The infite lap pool seamlessly blends its 50 feet of length with the mood of sky and ocean beyond. Pathways wind through the site following a historical wagon trail laid by California's early settlers.

这个山坡上 8 094 平方米梯田式的植物园在高处俯瞰着太平洋。它的特点是园里园外的远景，和两道现场挖掘的 100 英尺长的用作倾斜墙所使用的石头。50 英尺长的无边际泳池与天空和远处的海洋连成一色。园子里蜿蜒的小路是顺着加利福尼亚早期定居者留下的马车道修建的。

Its gardens are inspired by our client's mutual interest in varied regional gardens. This collection carefully weaves together a Woodland Garden, Provencal Garden, Japanese Zen Garden, Cacti Garden, Succulent Garden, California Native Garden, and other intercontinental low water using flora. The complex hillside construction includes the installation of numerous mature trees.

里面小花园的灵感来自我们客户对各式区域特色花园的共同兴趣。这里精心地汇集了一个特色花园：树林花园、普罗旺斯花园、日本禅园、仙人掌园、肉质植物园、加州原生花园，以及其他洲际的低水位植物。山坡上的建筑很复杂，栽有大量的成年树。

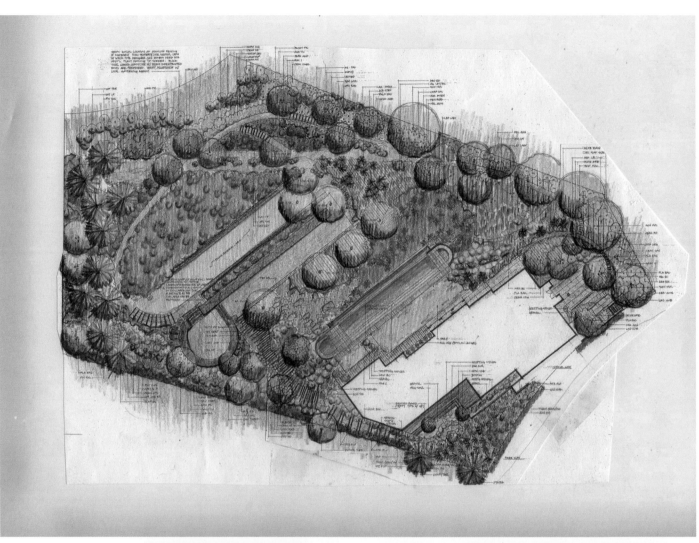

Trenton Drive

特伦顿车道花园

LOCATION：CA，USA
项目地点：美国 加利福尼亚

AREA：2,023 m²
面积：2 023 平方米

COMPLETION DATE：2010
完成时间：2010 年

PHOTOGRAPHER：John Feldman
摄影：John Feldman

ILLUSTRATIONS：John Feldman
插画：John Feldman

DESIGNER：Ecocentrix Landscape Architecture
设计师：Ecocentrix Landscape Architecture

DESIGN COMPANY：Ecocentrix
设计公司：Ecocentrix

Trenton Drive

We were charged with reinventing the gardens for this Colonial Revival style home in Beverly Hills, California. What was perhaps our greatest challenge was to orient a clean—lined, classical estate garden around an existing oddly shaped swimming pool. By incorporating herringbone and basket—weaved brick patterns, along with subtle inclusion of "X" patterns in other paved details and site accessories, we were able to fashion this pool into something of "focal piece". Clipped boxwood borders and tall Podocarpus gracillior hedges provide containment to other floral riches within the garden.

我们被委以重任为这个殖民地复兴风格的住所改造花园，它位于加利福尼亚州的贝弗利山上。我们可能面临的最大挑战是围绕一个已有的形状奇特的游泳池建一个线条简洁而古典的花园。我们用砖头铺出人字形和篮子编织图案，在其他边角地方精心地铺出 X 形图案，另外加了一些装饰物，就把这个水池塑造成一个焦点。边缘修剪整齐的黄杨树和高高的罗汉松树篱把花园里的其他花卉围了起来。

While classical by order, there is a distinct use of plant materials that within the boxwood—edged planter beds is still very "California" by nature. There is a purposeful overlay of contemporary space in the creation of this classical garden. Clean lines and distinct forms give strong cue to the orthogonal architecture. The "dogleg"—shaped pool was a hurdle that was overcome by strong design allegiance to symmetry and axial circulation through the garden.

按照要求，花园必须具有古典美，但在以黄杨树为边的花坛里使用的植物很特别，本质上仍是加利福尼亚风格。在创造这个古典的花园时，还特意添加了一层现代空间。简明的线条和清晰的外形强烈地体现着直角型建筑。外形参差不齐的水池是一个障碍，它通过设计上严格执行对称和轴心环绕的方案得以解决。

Casual crushed gravel pathways relent a wonderful sound under foot, and carry through to the dining garden. Brick paving details were used in other areas to create intended dialogue with lawn planted joints found in concrete paving and also to mimic the branded "X" on the black enamel painted Versailles planters.

碎卵石铺出的休闲小路使脚步变得轻盈，一直通到餐厅花园。在其他细枝末节的地方使用了砖铺，一是试着和混凝土地面上的草坪连接点相搭配，二是模仿具有黑色釉面的凡尔赛花盆上面的 X 标志。

Night lighted existing trees are majestic in this garden and equally so is the moonlight effect cast from high—mounted fixtures in their uppermost canopies. Path lighting is abundant yet subtly strikes a balanced painting and even washes across the more highly trafficked areas.

花园里夜光下的树十分迷人，放置在最顶部的灯抛洒下的光像月光一样，也是那样迷人。路的照明很充足，在交通流量较高的地方，洒下来的光像一幅色彩均匀的水彩画。

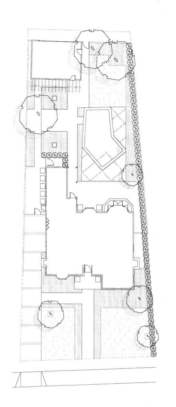

Iron privacy rails adorn the tops of white washed brick walls, while a trellis, fences, and gates, are boldly crafted from milled lumber. They all are fashioned to reflect and enhance the inherent details found in this classical Beverly Hills, residential architecture. The context of built and living details is congruent with its architecture and ground this home that is steeped in rich history and tradition.

铁制的民用栏杆装饰在白漆砖墙的顶部，格架、栏栅和大门明显是用细工木料精心制作的。它们的塑造反映和提升了这个经典贝弗利山住所建筑的内在细节。建筑和居住的细节要符合整个建筑风格，是体现这个有着丰富历史和传统之家的基石。

West Adams Residence

West Adams 住所

LOCATION: CA, USA
项目地点：美国 加利福尼亚

AREA: 1,012 m²
面积：1 012 平方米

PHOTOGRAPHER: John Feldman
摄影：John Feldman

DESIGNER: Ecocentrix Landscape Architecture
设计师：Ecocentrix Landscape Architecture

DESIGN COMPANY: Ecocentrix
设计公司：Ecocentrix

West Adams Residence

Our clients came to us with one wish for their garden which was inspired by their vision of French Parks. While the existing space was not as spatially expansive as most French Parks, we used elements such as a narrow 40 feet long reflecting pool, shaped hedges, and multiple sequenced outdoor rooms—let into through living archways of edible Bay Hedge plants. Pathways remained casual underfoot with the sparkle of crushed seashells mixed with decomposed granite.

客户带着一个愿望来找我们，希望他们的花园能按照他们头脑中法国公园的样子来设计。虽然花园在空间上没有大多数法国公园开阔，但我们使用了一些元素，比如，一个长40英尺的狭窄的倒影池，经过造型处理的树篱和多个排列有序的室外房间——从用可食用的海湾树篱植物构成的拱形门进入。里面的小路铺有闪烁光芒的碎贝壳和风化的花岗岩，踩在脚下让人感到十分放松。